A Guide to

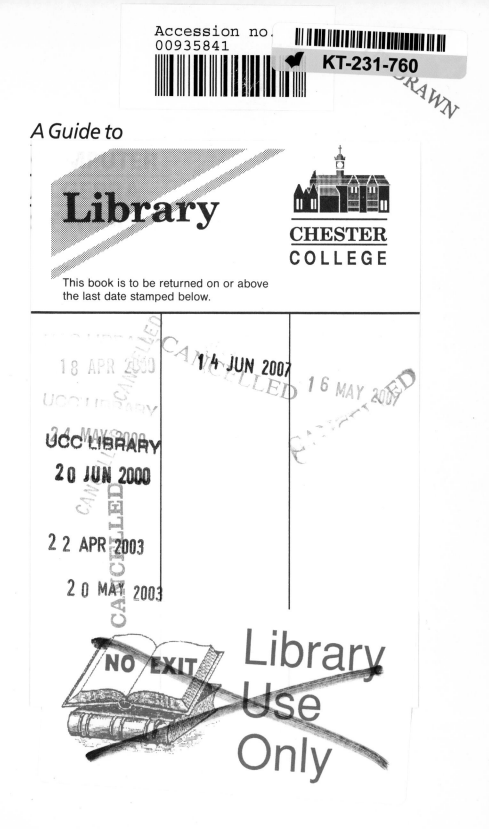

Library

CHESTER COLLEGE

A Guide to
COMPUTER
ALGEBRA
SYSTEMS

David Harper
Queen Mary & Westfield College, University of London

Chris Wooff
University of Liverpool

David Hodgkinson
University of Liverpool

JOHN WILEY & SONS
Chichester · New York · Brisbane · Toronto · Singapore

Other Wiley Editorial Offices

John Wiley & Sons, Inc., 605 Third Avenue,
New York, NY 10158-0012, USA

Jacaranda Wiley Ltd, G.P.O. Box 859, Brisbane,
Queensland 4001, Australia

John Wiley & Sons (Canada) Ltd. 5353 Dundas Road West,
Fourth Floor, Etobicoke, Ontario M9B 6H8, Canada

John Wiley & Sons (SEA) Pte Ltd, 37 Jalan Pemimpin 05-04,
Block B, Union Industrial Building, Singapore 2057

A catalogue record for this book is available from the British Library

ISBN 0 471 92910 7

Printed in Great Britain by Biddles Ltd, Guildford

Contents

Preface

This book is the first to combine in a single volume a comparison of the popular computer algebra systems. It had its origins in a booklet which was produced as part of the Computer Algebra Support Project at the University of Liverpool. The Project began in 1987 and its aim was to promote the use of computer algebra systems in the U.K. academic community. In the management of the Project, we were frequently asked: *What computer algebra systems are available? What facilities do they offer? What are their relative merits?*

To answer these questions, we wrote the first version of *A Guide to Computer Algebra Systems*. The booklet proved very popular and underwent several revisions as new computer algebra systems appeared and the existing systems were enhanced. The present book borrows much from the earlier *Guide*, both in format and in spirit. It is intended to be a gentle introduction to computer algebra rather than a treatise on how computer algebra systems are built. We hope that it will be useful to the reader who is intrigued by the idea of performing symbolic calculations with the aid of a computer and that it will also become a reference for anyone who needs a thorough and independent comparison of the five major systems when choosing the system which will meet their requirements most closely.

The book is arranged into three main parts: (i) *basic facilities* (ii) *case studies* and (iii) *bibliography*. Chapters 2 to 7 describe the facilities offered by computer algebra systems. Each chapter covers a general mathematical theme such as calculus or equation-solving, and the chapters are further sub-divided into specialised mathematical tasks. The chapter on calculus, for example, includes a description of the use of computer algebra systems for evaluating integrals algebraically. At the end of chapters 2 to 6, there are tables which summarise the abilities of the systems to perform the tasks described within the chapter. These tables provide an "at a glance" comparison of the five systems, enabling the reader to quickly assess the relevance of each system to their requirements. Chapter 8 illustrates the use of computer algebra systems to solve a selection of real problems – the type of mathematical problems which arise naturally in many areas of research in science and engineering. These case studies show how computer algebra systems are used in practice, and they highlight some of the differences between the various systems.

The final chapter is an annotated bibliography of books and papers on computer algebra. It will enable the reader to explore this rich and fascinating subject in greater depth.

The authors are indebted to Stan Schofield, Brian Denton, Victoria Pennington and Carl Murray for their careful reading and invaluable comments on the manuscript, and we are grateful for their sound advice.

David Harper
Queen Mary & Westfield College, University of London
Chris Wooff
University of Liverpool
David Hodgkinson
University of Liverpool

June 1991

Chapter 1
Introduction

THE ORIGINS OF COMPUTER ALGEBRA

It is often assumed that the idea of using computers to perform symbolic rather than numerical calculations is relatively new. In fact, computer algebra has had a very long childhood. The ideas date back to the days of the very first pioneers of computing. Lady Ada Lovelace, the patron of Charles Babbage first suggested symbolic computation or computer algebra a century and a half ago. However, it was not until the 1960s that Lady Lovelace's ideas could be realised. The following sections outline its subsequent developments.

SPECIALISED SYSTEMS

The first computer algebra systems were written to solve specific problems, usually in the field of Applied Mathematics. Many of them were non-interactive. Because the early systems were specific to one area of research and were relatively difficult to use, they made little or no impact in other fields of research. Many of these specialised systems were only available for a very limited number of operating systems, which further restricted their use.

GENERAL-PURPOSE SYSTEMS

The next generation of computer algebra systems were less specialised in that they were not written for the solution of problems in a single field. Some of the specialised systems did offer facilities required for the solution of a variety of problems. However, they still fell short of being truly general-purpose. The characteristics of the newer systems include:

- not restricted to a particular application

- better user interface

- wide range of built-in functions.
- can be used interactively
- available on a range of hardware platforms.

Largely because of these criteria, the number of general-purpose systems is much smaller. It includes the following packages: RE-DUCE, MACSYMA, Maple, SMP, Mathematica, muMATH, Derive and SCRATCHPAD.

The packages which are the main concern of this book are RE-DUCE, MACSYMA, Maple, Mathematica and Derive. However, the other three packages are also given brief mentions in the following sections.

The fact that all of these systems are general-purpose means that they can be used to great effect on a very wide range of applications. These systems are also more interactive than the specialised systems and give a much better user interface. These improvements have been responsible for the much more widespread use of computer algebra systems in both research and teaching.

REDUCE

REDUCE was the first general-purpose package. It became available in 1970. REDUCE was written by Anthony C. Hearn and is distributed by the RAND Corporation. It is one of the best known computer algebra systems and versions are available for a very wide range of operating systems.

REDUCE has relatively few built-in functions, but allows the user to define his/her own functions in a straightforward manner. REDUCE is currently the most widely used computer algebra system. This is one of the big advantages of using REDUCE: its widespread popularity ensures that a wide range of additional, user contributed packages are freely available.

MACSYMA

MACSYMA is a relatively large system which was developed at M.I.T. from 1969 until 1982. The development of MACSYMA was very much a team effort. MACSYMA first became commercially available in the mid 1970s. It is now marketed by Symbolics Inc. MACSYMA is available for fewer operating systems than REDUCE.

MACSYMA has a large number of built-in functions, which make it a very powerful system. MACSYMA and REDUCE are both written in the programming language LISP.

MAPLE

The development of Maple started at the University of Waterloo in 1980. Maple became generally available in 1983. WMSI (Waterloo Maple Software Inc.) was set up in 1989 specifically to market and distribute Maple.

Maple is presently used on a variety of mainframes, workstations and microcomputers. The Maple developers are still very active and each new version usually brings a significant number of enhancements.

SMP AND MATHEMATICA

SMP was written by Stephen Wolfram in the late 1970s and early 1980s. SMP has been largely superseded by Mathematica and it seems very unlikely that any new versions of SMP will be written.

Like MACSYMA, SMP has a fairly wide range of built-in functions. Versions of SMP existed for most of the popular operating systems.

Mathematica was launched amidst much excitement in 1988. Stephen Wolfram was once again the key developer. He set up Wolfram Research Inc., who are responsible for both the development and marketing of Mathematica.

Versions of Mathematica exist for a fairly wide range of computers. The wide range of graphics functions available with Mathematica are one of its greatest attributes.

muMATH AND DERIVE

muMATH development started in the late 1970s. It was the first computer algebra system which was written for microcomputers. David Stoutemyer and Albert Rich are the co-authors of muMATH and the co-owners of Soft Warehouse Inc., which markets both muMATH and Derive.

muMATH runs under the CP/M and DOS operating systems. In 1988, Soft Warehouse launched the Derive program. Derive is intended as a replacement for muMATH.

Derive runs under DOS. It is a menu-driven system which is extremely easy to use and is an ideal system for use in teaching.

SCRATCHPAD

SCRATCHPAD is currently an IBM research product and is not commercially available. It earns a brief mention here because it is a powerful system which is likely to become popular if it becomes widely available.

SCRATCHPAD has been actively developed by a group headed by Richard Jenks at IBM's T.J. Watson Research Center since the mid 1970s.

WHY USE COMPUTER ALGEBRA?

Now that the eight general-purpose systems have been introduced, it is opportune to consider some of the reasons for using computer algebra. Different users of computer algebra will give a variety of reasons for using these systems. In our experience, some of those reasons could be summarised as follows:

- Computer algebra can save both time and effort in solving a wide range of problems. Moreover, the solution is more likely to be correct than one obtained with more traditional techniques. Most exponents of computer algebra can cite examples in which a computer algebra system has quickly solved a problem which had been inhibiting a research activity.

- Algebraic solutions are often preferable to numerical ones. They are usually more compact than a set of numerical solutions. In addition, an algebraic solution can tell a researcher or student far more about the relationship between the variables in an expression than a host of figures.

- Algebraic solutions will always be exact. Numerical solutions will normally be approximations. This can arise from rounding and truncation errors which are inherent in the processing of real numbers. In addition, numerical data often needs to be presented in tabular form at some predetermined interval. This means that further errors can creep in when the user interpolates the data. When a computer algebra system is used to produce floating point results, then the accuracy is under the control of the user rather than the hardware.

- Computer algebra systems reduce the need for tables of functions, series and integrals. These tables are very popular with many researchers and are particularly heavily used by engineers and scientists. Computer algebra systems have highlighted many errors in such tables. Experts estimate that the error rate is approximately 15%.

- Currently, the teaching of applied mathematics has to involve much time in teaching techniques of solution. Computer algebra systems tend to produce solutions much more quickly and so they enable more time to be devoted to studying the properties of the solution.

- Computer algebra systems enable students to investigate the solution of larger problems than they could ever attempt using traditional methods.

- When a computer algebra system cannot find an exact solution or when a numerical solution is required, it is often useful to use the system to simplify expressions algebraically before evaluation.

- Computer algebra systems can be readily enhanced. Users can add routines which are relevant to their own applications.

Chapter 2

Basic Algebra

POLYNOMIAL ALGEBRA

The manipulation of multivariate polynomials is the most basic facility offered by computer algebra systems. They can add, subtract and multiply polynomials, cancelling or collecting like terms and extracting common factors where appropriate. Some of the earliest computer algebra systems were designed to automate the tedious and occasionally error-prone business of handling large polynomials. All modern computer algebra systems continue this tradition, and nobody with access to a computer algebra system would contemplate manipulating large polynomials by hand.

In addition to fundamental operations, computer algebra systems can re-write polynomials in a number of ways. A polynomial such as

$$x^3 - 8x^2 + 11x + 20$$

can be factorised to yield

$$(x - 5)(x - 4)(x + 1).$$

All of the computer algebra systems are able to factorise certain classes of polynomial. In general, they can factorise polynomials over the rationals, and some systems can factorise polynomials over complex or finite fields. However, factorisation of large polynomials often takes a very long time and is occasionally beyond the memory capacity of some computers.

Polynomials can be converted to Horner (nested) form. The example above becomes

$$\big((x - 8)x + 11\big)x + 20.$$

This form is the most efficient for numerical evaluation of a polynomial since it minimises the number of multiplications and eliminates explicit powers in the chosen variable.

The polynomial can also be re-written by collecting terms in powers of a chosen variable. Most computer algebra systems provide functions to determine the degree of a multivariate polynomial in any of its variables, and to extract the coefficient of a particular power of a variable. This enables polynomials to be analysed completely term-by-term.

ARITHMETIC

One of the features which distinguishes computer algebra systems from conventional programming languages (e.g. Pascal and FORTRAN) at an elementary level is the way in which they represent numbers. All five computer algebra systems are able to perform arithmetic on integers of arbitrary length. This enables us to calculate a number such as 200! in full with a single command. In Maple, for example, the command

```
200!;
```

results in the display

```
78865786736447905035523632139321850622951359776 8717
32632947425332443594499634033429203042840119846239
04177212138919638830257642790242637105061926624952
82993111346285727076331723739698894392244562145166
42402540332918641312274282948532775242424075739032
40321257405579568660226031904170324062351700858796
17892222278962370389737472000000000000000000000000
000000000000000000000000000
```

It would take substantial effort by a dedicated programmer to duplicate this result in FORTRAN, but it is a standard feature of computer algebra systems. The basic abilities to add, subtract and multiply integers of any length are extended by the ability to calculate the greatest common divisor of two integers in order to simplify fractions to their lowest form. By default, computer algebra systems preserve complete accuracy by using integers and ratios of integers (i.e. fractions) to represent numerical coefficients and indices within algebraic expressions.

In addition to arbitrary-precision integer arithmetic, computer algebra systems can also perform floating point arithmetic to any precision specified by the user. It is possible, for example, to carry out floating point calculations to 20 or 100 or 1000 digits using such systems. The command

```
N[Pi,40]
```

can be used to instruct Mathematica to calculate π to 40 digits, which yields

3.1415926535897932384626433832795028841971

Naturally, this type of calculation is done in software and so it is relatively slow. By contrast, floating point calculations in FOR-TRAN are normally performed in hardware and although the results are limited to a precision of about 16 digits, the computer can carry out millions of calculations per second. As a result, computer algebra systems can never replace languages such as FORTRAN for large-scale numerical calculations where 16 digits accuracy is sufficient. Instead, their arbitrary-precision floating point capability is useful when combining algebraic and numerical methods or when very high accuracy is essential.

USING PREVIOUS RESULTS

Algebraic calculations often proceed in a step-by-step manner, where each operation involves the results of the operations that immediately precede it. Thus it is often necessary to be able to recall the results of earlier calculations when using a computer algebra system. This can, of course, be achieved by assigning each result explicitly to a variable, as in a FORTRAN program, for example. However, all five computer algebra systems provide a means for recalling previous results very easily without the use of auxiliary variables. Most of the systems reserve a special symbol to denote the result of the previous calculation. This is WS in REDUCE, % in Mathematica and " in Maple. Thus

Factorise[%]

instructs Mathematica to factorise the result of the previous calculation.

Some systems also number each command or input expression and the corresponding result. The user can then recall any previous result during the session. Mathematica has the special function Out to enable users to retrieve earlier results, so that

    `Factorise[Out[23]]`

will factorise the result of the 23rd command issued during the
current Mathematica session.

Derive is extremely flexible in this respect. The user can scroll
forwards and backwards through the output of the current session
and select not only previous results but also individual terms and
sub-expressions by using the cursor keys to indicate the desired
term. This makes it relatively easy to perform such operations as
extracting the denominator of a previous calculation or isolating
selected terms of a series.

SPECIAL FUNCTIONS

Most mathematical calculations, especially those that arise from
problems in science and engineering, cannot be expressed solely
in terms of polynomials. Applied mathematics abounds with
special functions ranging from the familiar sine and cosine to
more specialised functions such as Airy functions and elliptic
integrals.

Each of the five computer algebra systems recognises a variety
of these functions together with some of their properties. These
are some of the functions that may be known to computer algebra
systems:

- Trigonometric functions and their inverses
- Logarithm and exponential functions
- Hyperbolic functions and their inverses
- Gamma and polygamma functions
- Error function
- Bessel functions
- Sine and cosine integral functions
- Families of orthogonal polynomials

and the properties may include

- Elementary simplification, e.g. $\sin(-x) = -\sin x$
- Values at special arguments, e.g. $\sin \pi = 0$
- Numerical evaluation, e.g. $\sin(0.1) = 0.09983361665$ to 10 significant figures
- Derivatives and integrals (see next chapter), e.g.

$$\frac{d}{dx} \sin x = \cos x$$

- Relationships between functions, e.g. $2 \sin x \cos x = \sin 2x$
- Inverse functions, e.g. $\sin(\arcsin x) = x$

The complete set of mathematical properties possessed by a function, or family of functions, is often very large, even for well-known functions such as sine and cosine. In addition, there are often many equivalent representations of the same expression, each of which is the "best" representation in a particular context. Thus it would be meaningless to make a computer algebra system apply all known properties of a function simultaneously. As a result, most computer algebra systems will only apply a very limited subset of such properties. Where necessary, the user can often define additional properties for special functions.

COMPLEX ALGEBRA

The imaginary unit $i = \sqrt{-1}$ appears frequently in mathematics, especially in fields such as physics and electrical engineering. All five computer algebra systems reserve a special symbol to denote i and they can perform elementary simplifications on complex expressions. The degree of simplification that can be achieved varies between systems. REDUCE merely regards i as an ordinary polynomial variable with the additional property that $i^2 = -1$. It cannot simplify expressions which include special functions with complex arguments such as $\sin(x + iy)$. Most of the other systems are a little more sophisticated. Maple, for example, can convert

$$\sin(x + iy)$$

into

$$\sin x \cosh y + i \cos x \sinh y$$

and can simplify rational expressions involving complex quantities.

The ability to separate real and imaginary parts of a complex expression is also quite limited in all of the systems: it generally extends no further than selecting those terms in an expression which contain i as an explicit algebraic factor. These terms are regarded as the imaginary part of the expression, and the remainder is the real part.

A true complex function environment is quite complicated and is the subject of active research.

HOW SIMPLIFICATION RULES
ARE APPLIED

During the course of a calculation, it is often necessary to apply different simplification rules at each stage. In some cases, the rules applied at one stage are the reverse of the rules applied at an earlier stage. For example, an expression containing sine and cosine terms may be converted to complex exponential notation at the start of a calculation in order to manipulate it as a polynomial. At the end of the calculation, it is then converted back into sine and cosine form. Clearly, a mechanism must be provided to enable the two transformations to be applied independently. They cannot both be in force at the same time, otherwise the computer algebra system would enter an endless loop applying them alternately.

Computer algebra systems employ several approaches to this problem. The first, exemplified by REDUCE, is to use *switches* or Boolean flags to indicate which rules are in effect at any instant.

For example, REDUCE has a switch called `FLOAT` which governs whether numbers in an expression are to be represented as exact rational fractions or approximate floating point numbers. The default value of `FLOAT` is *false*, causing REDUCE to use exact rational arithmetic. The command

    ON FLOAT;

sets the value of `FLOAT` to *true*, and REDUCE then evaluates any numerical constant in an expression as a floating point number. The switch can be set to *false* with the command

    OFF FLOAT;

whereupon REDUCE reverts to exact arithmetic once again.

Switches are an important part of REDUCE and they can be used to control many aspects of the system's behaviour, including the way in which expressions are displayed.

Derive also uses switches via its menu options. The user can specify the type of simplification to be applied to several types of function. For example, trigonometric functions can be simplified by *collecting*. This converts powers and products of sine and cosine into linear expressions with compound arguments. *Expanding* is the reverse transformation. Derive allows the user to choose either to *expand* or *collect* trigonometric terms.

Another approach is used by Maple, which has a variety of procedures that can be invoked to simplify an expression or to evaluate it in a particular form. For example, the `evalf` procedure evaluates its argument in floating point form whilst `radsimp` simplifies an expression which contains radicals. In addition, the `simplify` and `convert` procedures can be used to apply specific types of simplification or conversion to an expression. This approach has the advantage of being more selective than global rules controlled by switches. Each stage of the calculation can be controlled more closely. However, it frequently demands more work on the part of the user. Newcomers to systems such as Maple often have to experiment with the wide range of simplification procedures in order to discover which ones are the most useful under particular circumstances.

A third approach is used by Mathematica. A set of transformation rules can be specified explicitly at each stage of a calculation, to be applied only to the current expression. For example,

```
Sin[2a] + Sin[4b] /. Sin[2 x_] -> 2 Sin[x] Cos[x]
```

instructs Mathematica to apply the transformation rule which converts $\sin(2X)$ into $2\sin X \cos X$ for any X, yielding

```
2 Sin[a] Cos[a] + Sin[4b]
```

The `/.` symbol tells Mathematica to simplify the left-hand side according to the rules on the right-hand side.

Notice that Mathematica applies the rule to $\sin 2a$ but not to $\sin 4b$, since it did not recognise $4b$ as $2 \times 2b$.

DEFINING NEW RULES

There are three principal methods for adding new rules to a computer algebra system. They can be called the **declarative**, **procedural** and **imperative** methods. Four of the five computer algebra systems provide at least one of these methods; Derive does not allow new rules to be defined and so it is excluded from the present section.

The **declarative** method is used by REDUCE. It allows global rules to be declared. For example, to instruct REDUCE to use the substitution rule

$$\sec^2 x \to 1 + \tan^2 x$$

for all values of x, the user gives the command

```
FOR ALL X LET SEC(X)**2 = 1 + TAN(X)**2;
```

and this rule is applied during each subsequent calculation in addition to all of the rules already defined. The declaration is simply a statement telling REDUCE to replace one pattern ($\sec^2 X$, for any X) with another ($1 + \tan^2 X$). REDUCE does not check the mathematical veracity of any new rule. It could be given the command

```
FOR ALL X LET SEC(X)**2 = 1 + SIN(X)**2;
```

and it would change $\sec^2 X$ into $1 + \sin^2 X$ without question. Moreover, it does not check for contradictions between rules, so that although

```
FOR ALL X LET SEC(X)**2 = 1 + TAN(X)**2;
FOR ALL X LET TAN(X)**2 = SEC(X)**2 - 1;
```

might appear to be consistent declarations of the same mathematical identity, REDUCE will enter an infinite loop if asked to simplify, say, $\sec^2 x$ when both rules are active simultaneously.

The **procedural** method is used by Maple and is less straightforward and convenient than the declarative method. It requires the user to write a procedure in Maple's programming language to analyse an expression and to modify it to apply the desired rule. In order to apply the $\sec^2 X$ rule globally when using Maple, the procedure must search each part of an expression for occurrences of $\sec^2 X$ and explicitly replace them with $1 + \tan^2 X$.

Fortunately, Maple allows the user to analyse an expression term by term using the **whattype, nops** and **op** functions. Consider the expression

$$y = 6(1 + x) + \sec^2 p - a$$

Maple regards this as a sum of three terms, namely $6(1 + x)$, $\sec^2 p$ and $-a$ so the command

```
whattype(y);
```

yields

```
+
```

The **op** function can be used to select individual terms. The second term is obtained using the command

```
z := op(2,y);
```

which assigns the value $\sec^2 p$ to the variable z. Then

```
whattype(z);
```

yields

(the symbol for exponentiation) which indicates that the term is of the form a^b. The op function can be used again to extract the two parts of the expression:

```
op(1,z);
```

yields

```
sec(p)
```

and

```
op(2,z);
```

yields its exponent, 2. Further application of **whattype** and op establishes that the term is of the form $\sec^2 X$ for some X (p in this case) and so the $\sec^2 X \rightarrow 1 + \tan^2 X$ rule can be applied to it.

The **imperative** method is used by Mathematica, and a simple example was given in the previous section. It is possible to assign a name to a set of rules so that the set may be used repeatedly during a session. For example,

```
trigrules = {Sin[2 x_] -> 2 Sin[x] Cos[x],
             Cos[2 x_] -> Cos[x]^2 - Sin[x]^2}
```

gives the name **trigrules** to the specified rules. They can then be applied to any expression by giving a command such as

```
Sin[2 a] Cos[2 b] /. trigrules
```

Two or more sets of rules can be applied simultaneously with a command such as

```
Sin[2 a] Log[x^5] /. {trigrules, logrules}
```

but, as with REDUCE, the user must take care to avoid applying rules which are mutually contradictory.

NOTE ON TABLES

The following abbreviations are used in the tables in the next section: **RED** = REDUCE, **MPL** = Maple, **MAC** = MACSYMA, **DER** = Derive, **MTH** = Mathematica.

SUMMARY

Feature	RED	MPL	MAC	DER	MTH

Elementary algebra

	RED	MPL	MAC	DER	MTH
Polynomial expansion	Yes	Yes	Yes	Yes	Yes
Factorisation	Yes	Yes	Yes	Yes	Yes
Horner form	No	Yes	No	No	No
Get degree of polynomial	Yes	Yes	Yes	No	Yes
Extract coefficients	Yes	Yes	Yes	No	Yes
Re-write polynomial in powers of chosen variable	No	Yes	Yes	Yes	Yes

Arithmetic

	RED	MPL	MAC	DER	MTH
Arbitrary-precision integer arithmetic	Yes	Yes	Yes	Yes	Yes
Arbitrary-precision floating point arithmetic	Yes	Yes	Yes	Yes	Yes
Use of floating-point co-processor	No	Yes	No	No	Yes
Numerical evaluation of polynomials	Yes	Yes	Yes	Yes	Yes
Numerical evaluation of special functions	Yes	Yes	Yes	Yes	Yes
Numerical evaluation of special constants	Yes	Yes	Yes	Yes	Yes

Using previous results

	RED	MPL	MAC	DER	MTH
Recall previous result	Yes	Yes	Yes	Yes	Yes
Recall all previous results	Yes	Yes	Yes	Yes	Yes
Recall previous input	Yes	No	Yes	No	Yes
Recall all previous inputs	Yes	No	Yes	No	Yes

Feature	RED	MPL	MAC	DER	MTH

Special functions

Feature	RED	MPL	MAC	DER	MTH
Trigonometric functions	Yes	Yes	Yes	Yes	Yes
Natural logarithm	Yes	Yes	Yes	Yes	Yes
Logarithm to any base	No	Yes	Yes	Yes	Yes
Exponential	Yes	Yes	Yes	Yes	Yes
Hyperbolic functions	Yes	Yes	Yes	Yes	Yes
Gamma function	No	Yes	Yes	Yes	Yes
Polygamma functions	No	Yes	Yes	Yes	Yes
Error function	Yes	Yes	Yes	Yes	Yes
Bessel functions of integer order	No	Yes	Yes	Yes	Yes
Sine and cosine integral functions	No	Yes	Yes	Yes	Yes
Elliptic integrals	No	Yes	Yes	Yes	Yes
Hypergeometric functions	No	Yes	Yes	Yes	Yes
Airy functions	No	Yes	No	Yes	Yes
Orthogonal polynomials	No	Yes	Yes	Yes	Yes

Complex algebra

Feature	RED	MPL	MAC	DER	MTH
Simplification of $i^2 \rightarrow -1$	Yes	Yes	Yes	Yes	Yes
Simplification of $(a + ib)/(c + id)$	No	Yes	Yes	Yes	Yes
Separation of real and imaginary parts	No	Yes	Yes	Yes	Yes
Special functions of complex arguments	No	Yes	Yes	Yes	Yes

Simplification rules

Feature	RED	MPL	MAC	DER	MTH
Controlled by flags	Yes	No	Yes	Yes	No
Simplification functions	No	Yes	Yes	No	Yes
Explicit transformation rules	No	No	No	No	Yes

Feature	RED	MPL	MAC	DER	MTH

Defining new rules

	RED	MPL	MAC	DER	MTH
Declarative	Yes	No	Yes	No	Yes
Procedural	No	Yes	Yes	No	Yes
Imperative	No	No	No	No	Yes

Chapter 3
Calculus

DIFFERENTIATION

After elementary polynomial algebra, the techniques of symbolic differentiation are probably the easiest mathematical operations to automate since they are governed by a handful of simple rules. The application of these rules will always yield an answer. Hence computer algebra systems can differentiate any expression involving functions whose derivatives are known to the system. They apply the rules for differentiation of products and quotients as well as the "chain rule" which defines the derivative of a composition of functions.

When differentiating a function of two or more variables, computer algebra systems evaluate the partial derivative. For example, an expression such as

$$a \sin(bx + c)$$

can be differentiated with respect to x to yield

$$ab \cos(bx + c)$$

or with respect to b to yield

$$ax \cos(bx + c)$$

or with respect to a or c. The computer algebra system regards all variables within an expression as valid variables for differentiation.

Most computer algebra systems offer a compact and convenient notation for calculating mixed and higher derivatives. For example, the Maple command

```
diff(F, x);
```

evaluates the single (partial) derivative

$$\frac{\partial F}{\partial x}$$

whilst the command

```
diff(F, x$2, y);
```

evaluates the mixed partial derivative

$$\frac{\partial^3 F}{\partial x^2 \partial y}.$$

It is often necessary to calculate the derivatives of functions whose dependence on their arguments is not explicitly stated: a function F may depend on x and y in a way which the user either does not know in advance or does not wish to specify. REDUCE and MACSYMA allow the user to make a declaration such as

```
DEPENDS F,X,Y;
```

which states that F depends upon x and y, and so the partial derivatives $\partial F/\partial x$ and $\partial F/\partial y$ should not be assumed to be zero.

All five systems allow dependencies to be declared using generalised operator (or function) notation. Thus $F(x, y)$ is explicitly dependent upon x and y, although the properties of F need not have been stated. REDUCE requires the user to declare that the name F is to be used as an operator or function with the command

```
OPERATOR F;
```

whilst the other computer algebra systems allow F to be used as an operator without any prior declaration.

Rules for calculating the derivative of a user-defined function can easily be defined using most of the computer algebra systems. In REDUCE, for example, the user can declare F and G to be operators with the command

```
OPERATOR F,G;
```

and then declare that $G(x) = dF(x)/dx$:

```
FOR ALL X,Y LET DF(F(Y),X)=DF(Y,X)*G(Y);
```

whilst in Maple, the same result would be achieved by defining a procedure called `diff/F` which would be invoked automatically by `diff` when required to evaluate the derivative of F. A command such as

```
diff(F(y), x);
```

would result in Maple invoking a procedure call

```
'diff/F'(y, x);
```

(the quotation marks around `diff/F` denote that the name of the function includes a slash (/), and that this does not indicate division).

Some computer algebra systems cannot deal with implicit dependencies such as

$$F(x, y(x))$$

since they do not make a distinction between the partial derivative $\partial F/\partial x$ which is the derivative with respect to the first argument, and the total derivative dF/dx which is

$$\frac{\partial F}{\partial x} + \frac{dy}{dx}\frac{\partial F}{\partial y}$$

Wester and Steinberg (1984) have shown that MACSYMA, REDUCE and early versions of Maple give incorrect results in such circumstances because they do not possess a notation which can represent the two types of derivative in an unambiguous way.

Mathematica and recent versions of Maple recognise the concept of differentiation as an operator acting upon a function. They both possess operators which provide an unambiguous notation for derivatives. Using Mathematica, the partial derivative $\partial F/\partial x$ in the example above would be written as

```
Derivative[1,0][F][x,y[x]]
```

whilst the total derivative dF/dx would be written as

```
Derivative[1,0][F][x,y[x]] +
   D[y[x],x] Derivative[0,1][F][x,y[x]]
```

Mathematica's `Derivative` and Maple's D are operators upon functions. The Mathematica notation

```
Derivative[m,n][F]
```

represents a function which is the mth derivative of F with respect to its first argument and the nth derivative with respect to the second. It can be used as if it were a function name itself.

INTEGRATION

The integration of algebraic expressions is one of the main applications of computer algebra systems: many problems in science and engineering require the evaluation of integrals. These may be indefinite, in which case they can be regarded as anti-derivatives, or definite, where they may represent the area or volume of a figure, or another of its bulk properties such as mass or moment of inertia.

Each of the five computer algebra systems can integrate a variety of expressions. Each system has a repertoire which always includes polynomials and simple transcendental functions. This may extend to include a wider class of integrable expressions depending upon the integration algorithms within the computer algebra system. All of the computer algebra systems employ familiar techniques such as substitution and "look-up tables" of known integrals. They also use more sophisticated methods based upon the Risch-Norman algorithm and the work of Bronstein, Trager and Davenport. It is almost impossible to put the five computer algebra systems into a "league table" according to their ability to evaluate indefinite integrals, so instead, several example integrals were devised and presented to the systems for evaluation. The results are summarised in the table at the end of the chapter.

The evaluation of definite integrals may be carried out in a naïve way by first obtaining the indefinite integral and then substituting the limits of integration. This method can be employed when using any of the five systems, providing that the system is able to determine the indefinite integral. However, there are many cases where the definite integral may be expressed in exact algebraic form even when the indefinite integral cannot. A good example is:

$$\int_0^\infty e^{-(ax)^2} dx = \frac{\sqrt{\pi}}{2|a|}$$

The corresponding indefinite integral,

$$\int e^{-(ax)^2} dx$$

cannot be written in exact form, yet some computer algebra systems are able to determine the definite integral using other methods.

There are also cases where simple substitution of the limits of integration into the indefinite integral is inappropriate. This may occur when the indefinite integral is unbounded at some point within the region of integration. For example,

$$\int_1^3 \frac{dx}{(x-2)^2}$$

is not convergent. However, most computer algebra systems will return a finite value for this integral. They do not check the integrand for unboundedness or continuity, and hence give a spurious answer. The exception is MACSYMA, which correctly reports that the integral is not convergent. In some cases, MACSYMA will enquire about the sign of algebraic variables within a definite integral in order to try to determine whether the integral is convergent.

When a definite integral has a purely numerical value, such as

$$\int_0^\pi \sin(\sin x) \, dx$$

then numerical methods (quadrature) may be used to obtain an approximate result. Some computer algebra systems include numerical quadrature as an extension of their methods of evaluating definite integrals.

So far, it has been assumed that the integral is a real function of a real variable. In applied mathematics, it is often necessary to integrate a complex function along a path in the complex plane, an operation usually denoted by

$$\oint f(z)dz$$

Such integrals arise in the branch of complex analysis known as the calculus of residues. A major part of the art of this subject lies in selecting an appropriate path along which to evaluate the integral. It may be necessary, for example, to take a geometrically regular path such as a straight line and to deform it to include or exclude singular points of the integrand function.

At present, computer algebra systems offer no facilities for performing complex contour integration directly. This is an area in which much useful development work could be done.

DIFFERENTIAL EQUATIONS

Many problems in science and engineering can be reduced to differential equations. The ability to solve differential equations is, therefore, of great importance in applied mathematics. One of the questions which the authors are most frequently asked about computer algebra systems is: *Can they solve differential equations?*

MACSYMA and Maple are capable of solving a wide range of first-order ordinary differential equations (ODEs) including nonlinear differential equations such as

$$\left(\frac{dy}{dx}\right)^2 - y^2 - 2xy = x^2$$

and they can also solve some of the simpler types of higher-order

ODE such as linear inhomogeneous equations. However, they are unable to recognise and solve differential equations such as

$$\frac{d^2 f}{dx^2} + (a - 2q\cos x)f = 0$$

which are non-linear.

They can solve certain types of simultaneous ODEs using techniques such as Laplace transforms where appropriate, and they can also calculate approximate series solutions to a variety of ODEs. Nevertheless, much work remains to be done in developing algorithms for solving differential equations using computer algebra systems.

None of the computer algebra systems can solve any class of partial differential equations, and this is one of the areas in which there is great potential for development.

LIMITS

It is often necessary to know the behaviour of a function as its argument approaches certain critical values, notably zero and infinity. Some of the five computer algebra systems can calculate the limit of an expression as one of the variables approaches a chosen value. This is quite distinct from simply substituting the value in place of the variable, of course, since that might lead to meaningless expressions such as $0/0$ or ∞/∞. The computer algebra systems use a number of techniques for evaluating limits, including L'Hôpital's rule and substitution of $1/\xi$ for x to replace a $x \to \infty$ limit by a $\xi \to 0$ limit. Most of the systems also allow a limit to be approached from either direction. For example, the limit of

$$\frac{1}{x - a}$$

is $+\infty$ if $x \to a$ from above and $-\infty$ if $x \to a$ from below.

SUMS AND PRODUCTS OF SEQUENCES

All of the five computer algebra systems are able to sum sequences over a finite range whose bounds are explicitly specified as integers. For example, they can all evaluate

$$\sum_{n=1}^{100} n^{20}$$

Some systems also include more sophisticated methods for evaluating sums over indefinite ranges enabling the user to calculate sums such as

$$\sum_{n=1}^{N} n^k$$

or over infinite ranges, to simplify

$$\sum_{n=0}^{\infty} \frac{x^n}{n!}.$$

However, they can only apply elementary simplifications to general sums. None of them can convert an expression such as

$$\sum_{n=2}^{\infty} n(n-1)a_n x^n + \sum_{n=1}^{\infty} na_n x^n + (x^2 - \nu^2)\sum_{n=0}^{\infty} a_n x^n$$

(which arises from the solution of Bessel's equation by power series) into

$$-\nu^2 a_0 + (1 - \nu^2)a_1 x + \sum_{n=2}^{\infty} \left((n^2 - \nu^2)a_n + a_{n-2}\right) x^n.$$

Thus, without substantial programming, they are of limited use, for example, in solving differential equations by general power series methods. Even their ability to evaluate indefinite sums is incomplete, and this is another area where improvements can be made.

Products of sequences appear less often than sums, and most of the systems can only evaluate products over specified integer ranges.

TAYLOR SERIES EXPANSIONS

Most of the five systems can calculate the Taylor series expansion of an expression about any point to a specified number of terms. This is useful for approximating a function in the vicinity of a point, and also for investigating the behaviour of the function near that point. For example, the function

$$\sin \tan x - \tan \sin x$$

has a Taylor series whose first few terms are

$$-\frac{1}{30}x^7 - \frac{29}{756}x^9 + \cdots$$

and this result can be calculated using the Maple command

```
taylor(sin(tan(x))-tan(sin(x)), x=0, 10);
```

Some of the systems are also able to calculate Laurent expansions of functions whose behaviour is asymptotic about a particular point.

REDUCE is the only system which does not have a built-in Taylor series function, but it is very easy to define a REDUCE procedure to evaluate Taylor series expansions.

INTEGRAL TRANSFORMS

Integral transforms such as Laplace and Fourier transforms provide powerful techniques for solving problems in many areas of applied mathematics. For example, Fourier transforms find many applications in theoretical physics and engineering, and Laplace transforms are widely used to solve differential equations.

Most of the five computer algebra systems are able to calculate the Laplace transform of an expression, either by evaluating the transform

$$f(s) = \mathcal{L}(F(t)) = \int_0^\infty F(t)e^{-st}dt$$

directly as a definite integral or by resolving the expression into several terms whose transforms may be found in a look-up table.

The calculation of inverse Laplace transform is a more difficult task. The inverse transform of a general function $f(s)$ is given by the Bromwich integral,

$$F(t) = \mathcal{L}^{-1}(f(s)) = \frac{1}{2\pi i} \int_{\gamma-i\infty}^{\gamma+i\infty} e^{st} f(s) ds$$

which is an integral in the complex plane. Computer algebra systems cannot evaluate such integrals and so they cannot calculate inverse Laplace transforms by this method. They must rely instead upon a comprehensive look-up table of inverse transforms. Not all of the systems which can calculate Laplace transforms are able to calculate inverse transforms.

NOTE ON TABLES

The following abbreviations are used in the tables in the next section: **RED** = REDUCE, **MPL** = Maple, **MAC** = MACSYMA, **DER** = Derive, **MTH** = Mathematica.

SUMMARY

Feature	RED	MPL	MAC	DER	MTH
Differentiation					
Derivatives of polynomials	Yes	Yes	Yes	Yes	Yes
Derivatives of special functions	Yes	Yes	Yes	Yes	Yes
Higher and mixed derivatives	Yes	Yes	Yes	Yes	Yes
Implicit dependency	Yes	No	Yes	Yes	No
Dependency via operators	Yes	Yes	Yes	Yes	Yes
Declaration of derivative of user's function	Yes	Yes	Yes	No	Yes
Distinction between total and partial derivatives	No	Yes	No	No	Yes
Integration					
Indefinite integrals					
$\int (1+x)^6 \, dx$	Yes	Yes	Yes	Yes	Yes
$\int \sqrt{1-x+x^2} \, dx$	Yes	Yes	Yes	Yes	Yes
$\int \dfrac{1}{\sqrt{1-x+x^2}} \, dx$	Yes	Yes	Yes	Yes	Yes
$\int \sin^8 x \, dx$	Yes	Yes	Yes	Yes	Yes
$\int \sin^{-1} x \, dx$	Yes	Yes	Yes	Yes	Yes
$\int x^2 \log x \, dx$	Yes	Yes	Yes	Yes	Yes
$\int \dfrac{\sin x}{x} \, dx$	No	Yes	Yes	No	Yes
$\int \dfrac{\sin x}{x^2+5} \, dx$	No	No	No	No	No

Feature	RED	MPL	MAC	DER	MTH

Integration (continued)

Definite integrals					
$\displaystyle\int_0^\pi \sin^4 x\, dx$	Yes	Yes	Yes	Yes	Yes
$\displaystyle\int_0^1 \frac{dx}{\sqrt{1-x^2}}$	No	Yes	Yes	Yes	Yes
$\displaystyle\int_0^\infty e^{-x^2}\, dx$	No	Yes	Yes	Yes	Yes
$\displaystyle\int_{-2}^2 \frac{dx}{(x-1)^2}$	No	Yes	Yes	No	No
$\displaystyle\int_0^\infty \frac{dx}{x^2+a^2}$	No	Yes	Yes	Yes	Yes
Numerical integration					
$\displaystyle\int_0^\pi \sin \sin x\, dx$	No	Yes	Yes	Yes	Yes
$\displaystyle\int_0^\infty e^{-x^3}\, dx$	No	Yes	Yes	Yes	Yes

Differential equations

ODEs					
1st-order	No	Yes	Yes	Yes	Yes
2nd-order linear	No	Yes	Yes	Yes	Yes
2nd-order non-linear	No	Yes	No	Yes	No
Higher-order	No	No	No	No	No
Solution by series approximation	No	Yes	Yes	No	No
Solution by general power series	No	No	No	No	No
Solution by Laplace transforms	No	Yes	Yes	No	No
PDEs					
All types	No	No	No	No	No

Feature	RED	MPL	MAC	DER	MTH
Limits					
$\lim\limits_{x \to 0} \dfrac{\sin x}{x}$	No	Yes	Yes	Yes	Yes
$\lim\limits_{x \to \infty} \tan^{-1} x$	No	Yes	Yes	Yes	Yes
$\lim\limits_{x \to \infty} \sin x$	No	Yes	Yes	No	Yes
Sums and products					
Sums over a finite integer range	Yes	Yes	Yes	Yes	Yes
Sums over an indefinite range	No	Yes	Yes	Yes	No
Sums over an infinite range	No	Yes	Yes	Yes	No
Simplification of general sums	No	No	No	No	No
Products over a finite integer range	Yes	Yes	Yes	Yes	Yes
Products over an indefinite range	No	No	No	No	No
Products over an infinite range	No	No	No	No	No
Taylor series expansions					
Taylor and Maclaurin expansions	No	Yes	Yes	Yes	Yes
Integral transforms					
Laplace transforms	Yes	Yes	Yes	Yes	Yes
Inverse Laplace transforms	Yes	Yes	Yes	No	No
Fourier transforms	No	No	Yes	No	No
Inverse Fourier transforms	No	No	Yes	No	No

Chapter 4

Solving Algebraic Equations

FINDING ROOTS OF POLYNOMIALS

It is an easy matter to solve quadratic equations since there exists a simple formula which expresses the roots in terms of the coefficients of the equation. Solving cubics and higher-degree equations is not such an easy task. Although there are methods for solving cubics and quartics, these are difficult and time-consuming to apply. Furthermore, no such methods are available for solving general polynomials of degree 5 or higher.

Computer algebra systems provide a valuable tool for tackling the complicated mathematical task of solving polynomial equations. A command such as

```
solve(a*x^4 + b*x^3 +c*x^2 + d*x +e = 0, x);
```

will yield the four solutions of the general quartic when using any of the five computer algebra systems. All five systems can solve polynomial equations up to degree 4, whether the coefficients are entirely algebraic, entirely numerical or a combination of numbers and algebraic expressions. The solutions are given as exact expressions, even when they are purely numerical. For example, the polynomial equation

$$x^3 - 2x^2 + x - 1 = 0$$

can be solved to yield a set of three solutions including

$$\left(\frac{25}{54} + \frac{\sqrt{69}}{18}\right)^{1/3} + \left(\frac{25}{54} - \frac{\sqrt{69}}{18}\right)^{1/3} + \frac{2}{3}$$

and this may easily be evaluated to, say, 12 digits to give the approximate solution

$$1.75487766624$$

This result was obtained using just *two* commands to a computer algebra system. It would take much more work on the part of the user to achieve the same result using a language such as FORTRAN, even in conjunction with a numerical subroutine library such as the NAG library.

Nevertheless, the solutions to polynomial equations can often be very complicated: the four solutions of the general quartic would cover many pages of this book, if they were to be printed in full. Even quadratic equations may give rise to solutions that cannot easily be manipulated or simplified by a computer algebra system. Cubics and quartics yield solutions that contain nested radicals, and these are notoriously difficult to simplify.

Needless to say, computer algebra systems cannot find solutions to the majority of quintics or polynomial equations of higher-degree, since general solutions do not exist to equations higher than quartics. Computer algebra systems can solve certain types of higher-degree polynomial equations, notably those that have rational roots, but some computer algebra systems will not even attempt to solve a polynomial equation of degree greater than 5 if it contains coefficients that are not purely numerical.

Some computer algebra systems provide facilities for solving equations numerically, though the user must usually specify an interval within which the system can find a root. Computer algebra systems which enable the user to plot a graph of the function are especially useful in such cases, since a graph of the function enables the user to locate the approximate roots of the function very easily.

SIMULTANEOUS LINEAR EQUATIONS

In addition to solving a single equation in one unknown, all five computer algebra systems are able to solve certain classes of simultaneous equations in several unknowns. The simplest class, for which a solution may always be found, is that of N independent linear equations in N unknowns. This is related to inversion of an $N \times N$ matrix, but simultaneous linear equations can be solved directly using the same command that is used to solve a single equation. In Maple, for example, the command

```
solve({3*x+y=1, x-y-z=4, x+2*y+3*z=8},
       {x,y,z});
```

instructs the system to solve the three equations

$$3x + y = 1$$
$$x - y - z = 4$$
$$x + 2y + 3z = 8$$

for the unknowns x, y and z, yielding the result

$$x = 3, \; y = -8, \; z = 7$$

SIMULTANEOUS NON-LINEAR EQUATIONS

It is not possible, in general, to solve sets of non-linear polynomial equations, but some computer algebra systems provide facilities for attempting to find partial solutions or to generate equivalent sets of equations which may lead to solutions more easily.

Maple, for example, may return implicit solutions which are expressed in terms of the roots of other polynomial equations. When told to solve the two equations

$$x + y = 5$$
$$x^3 + y^2 = 11$$

it returns solutions expressed as functions of the roots of the auxiliary cubic equation

$$\zeta^3 - 16\zeta^2 + 75\zeta - 114 = 0$$

In this case, explicit solutions can then be found by solving the polynomial in ζ, but the auxiliary polynomial is often of degree 5 or higher and so exact algebraic solutions cannot be found.

Some computer algebra systems incorporate a command to reduce a set of simultaneous non-linear polynomial equations by computing the Gröbner basis of the set. In simple terms, the Gröbner basis is a set of equations which is mathematically equivalent to the original set. It is obtained by eliminating variables between equations, in much the same way that Gaussian elimination is applied to simultaneous *linear* equations. The Gröbner basis may be larger than the original set, but the solution set of the basis includes the solution set of the original equations.

Some members of the basis may be univariate, and hence soluble (perhaps only numerically, of course). This may enable the entire set to be solved. One of the Gröbner bases of the example above is

$$y^3 - 16y^2 + 75y - 114 = 0$$

$$x + y - 5 = 0$$

whilst another is

$$x^3 + x^2 - 10x + 14 = 0$$

$$x + y - 5 = 0$$

In both cases, the first equation contains only one of the unknowns. The roots of this equation may be substituted into the second to obtain the value of the other unknown.

NOTE ON TABLES

The following abbreviations are used in the tables in the next section: **RED** = REDUCE, **MPL** = Maple, **MAC** = MACSYMA, **DER** = Derive, **MTH** = Mathematica.

SUMMARY

Feature	RED	MPL	MAC	DER	MTH
Roots of polynomials					
Exact roots of quadratics, cubics and quartics	Yes	Yes	Yes	Yes	Yes
Approximate numerical roots	No	Yes	Yes	Yes	Yes
Location of roots within an interval	No	Yes	Yes	Yes	Yes
Graphing of functions	No	Yes	Yes	Yes	Yes
Simultaneous linear equations					
Exact solutions	Yes	Yes	Yes	Yes	Yes
Solution by least squares	No	Yes	Yes	Yes	Yes
Simultaneous non-linear polynomial equations					
Explicit solutions	No	No	Yes	No	Yes
Roots in terms of auxiliary equations	No	Yes	No	No	No
Gröbner bases	Yes	Yes	Yes	No	No

Chapter 5

Matrix and Vector Algebra

HOW MATRICES ARE REPRESENTED

All five computer algebra systems allow the user to create and manipulate arrays and matrices. They provide commands to declare the dimensions of a matrix and to access its components in the same way as conventional programming languages, but the components can be algebraic expressions. For example, the matrix

$$\begin{pmatrix} x^2 + a & -b \\ b & x - c \end{pmatrix}$$

can be declared and initialised in REDUCE with the commands

```
MATRIX M;
M := MAT((X^2+A, -B), (B, X-C));
```

or in Mathematica

```
M = {{x^2+a, -b}, {b, x-c}}
```

and the $(1,2)$ component can be accessed and assigned to the variable p with a command such as

```
P := M(1,2);
```

in REDUCE, or

```
p = M[[1,2]]
```

in Mathematica.

Maple allows the user to declare whether the matrix is of a special type. It recognises sparse matrices and symmetric matrices and it can store such matrices more efficiently by making use of their sparseness or symmetry properties. For example, only the upper triangle of a symmetric matrix needs to be specified, since Maple knows that the (i,j) and (j,i) components are identical. When working with sparse matrices, Maple will only store those components that the user has declared explicitly to be non-zero.

ELEMENTARY MATRIX ALGEBRA

Each of the five computer algebra systems provides syntax or
a set of procedures for performing basic matrix algebra such as
addition, multiplication and exponentiation. In REDUCE, for
example, if A, B and C have been declared to be 3×3 matrices
and the components of A and B have been assigned algebraic
values, then C can be set equal to the matrix sum of A and B
using the command

```
C := A + B;
```

or it can be set equal to the matrix product $A.B$ using the command

```
C := A * B;
```

The other four systems use a special symbol to denote matrix
multiplication, since the scalar multiplication operator * is com-
mutative. Thus to multiply two matrices in Mathematica, the
command would be

```
c := a . b
```

Matrix algebra in Maple is carried out using the `evalm` function, and the matrix multiplication operator is `&*` so that the
matrix product would be evaluated with the command

```
c := evalm(a &* b);
```

All five systems can evaluate sums, products and powers of
matrices, provided that the components of all the matrices have
been declared. Some of the systems, notably REDUCE, can also
evaluate expressions containing objects which represent general
matrices. This can be achieved by declaring a non-commutative
multiplication operator and using it within matrix calculations.

Some of the systems provide facilities for more general linear algebra calculations such as determining the rank of a matrix or the kernel of a linear transformation defined by a matrix. They can also create special matrices such as the Hessian matrix of a function with respect to its variables, or the Hilbert matrix

$H_n(x)$ whose (i, j) component is $1/(i + j - x)$.

MATRIX INVERSION

The five computer algebra systems are able to invert square matrices whose components are either numerical or algebraic. The command to invert a matrix may be provided as a function, such as Mathematica's Invert function, or it may be part of the general matrix algebra notation. To invert a square matrix A using REDUCE, for example, the user need only type

```
1/A;
```

The inverses of matrices with algebraic components are generally very complicated expressions, since every component of the inverse contains the determinant of the original matrix in its denominator. Unfortunately, most of the systems cannot factorise the matrix to remove the common denominator. The components of the inverse grow rapidly in complexity as the size of the matrix increases – the determinant of an $N \times N$ matrix has $N!$ distinct terms – and so attempts to invert large matrices with algebraic components may fail because the computer has insufficient memory.

Matrix inversion is also a time-consuming process, and it is unlikely that computer algebra systems will replace numerical routines such as those in the NAG or IMSL libraries as tools for inversion of large matrices whose components are purely numerical.

DETERMINANTS AND EIGENVALUES

In many mathematical problems arising in physics and engineering, it is necessary to determine the eigenvalues of a matrix. Computer algebra systems can calculate the determinant of a square matrix and can also solve algebraic equations, so they provide a useful tool for eigenvalue calculations. Most of the five computer algebra systems have commands to determine the eigenvalues and eigenvectors of a matrix. Using Mathematica, for example, the eigenvalues and eigenvectors of a general 2×2 matrix can be calculated with the command

```
{Mvals, Mvecs} = Eigensystem[{{a,b}, {c,d}}]
```

which assigns the eigenvalues to the variable `Mvals` and the corresponding eigenvectors to the variable `Mvecs`.

If the determinant is required explicitly, computer algebra systems possess functions to calculate it with a command such as

```
Det[{{a,b}, {c,d}}]
```

VECTOR AND TENSOR CALCULUS

All five computer algebra systems allow the user to perform vector algebra in three dimensions using the notation for matrix algebra. Vectors are simply regarded as column matrices. However, there are a number of operators which can only be applied to vectors. These are the vector cross product ($\times$) and the vector differential operators *div* and *curl*. The *grad* operator acts upon a scalar field to yield a vector field, and so it may also be considered in this context. Some of the systems provide a facility for performing vector calculus in any orthogonal curvilinear coordinate system whose coordinates and scale factors are defined by the user. Such coordinate systems arise in many branches of physics and engineering, including fluid mechanics and the study of electromagnetic fields.

Tensor algebra and calculus in an arbitrary number of dimensions is a notoriously difficult subject. The manipulation of tensors is frequently time-consuming and tedious since it requires a great deal of algebra and differentiation. Specialised computer algebra systems have been written to automate tensor calculations. However, the five systems described here are only able to carry out a limited range of operations on tensors. They are also unable to display tensors easily.

NOTE ON TABLES

The following abbreviations are used in the tables in the next section: **RED** = REDUCE, **MPL** = Maple, **MAC** = MACSYMA, **DER** = Derive, **MTH** = Mathematica.

SUMMARY

Feature	RED	MPL	MAC	DER	MTH
How matrices are represented					
Matrices must be pre-declared	Yes	No	No	No	No
Components must be fully specified	Yes	No	No	No	No
Components may be altered individually	Yes	Yes	Yes	Yes	No
Sparse matrices are recognised	No	Yes	No	No	No
Symmetric matrices are recognised	No	Yes	No	No	No
Elementary matrix algebra					
Matrix addition	Yes	Yes	Yes	Yes	Yes
Matrix multiplication	Yes	Yes	Yes	Yes	Yes
Scalar multiplication	Yes	Yes	Yes	Yes	Yes
Exponentiation	Yes	Yes	Yes	Yes	Yes
Special matrices are recognised	No	Yes	Yes	Yes	No
Matrix inversion					
Inversion of algebraic matrices	Yes	Yes	Yes	Yes	Yes
Determinants and eigenvalues					
Determinants	Yes	Yes	Yes	Yes	Yes
Eigenvalues	Yes	Yes	Yes	Yes	Yes
Eigenvectors	Yes	Yes	Yes	Yes	Yes
Characteristic equations	Yes	Yes	Yes	Yes	Yes

Feature	RED	MPL	MAC	DER	MTH

Vector and tensor calculus

	RED	MPL	MAC	DER	MTH
Vector algebra	Yes	Yes	Yes	Yes	Yes
Vector calculus in Cartesian coordinates	No	Yes	Yes	Yes	Yes
Vector calculus in general orthogonal curvilinear coordinates	No	No	Yes	No	No
Tensor algebra and calculus	No	No	Yes	Yes	No

Chapter 6

Input and Output

PRETTY-PRINTING

When computer algebra systems try to display the results of a calculation, they are faced with a problem: most computer terminals cannot print output in true mathematical notation, using superscripts and special symbols such as integral and summation signs. Most VDUs are character-based and so they can only show plain text. This has led to the use of a style known as "pretty-printing", which is an attempt to display the two-dimensional notation of mathematics on an ordinary VDU. It works by simply taking two (or more) lines of the VDU screen to show one line of mathematics. For example, the expression

$$x^4 - 3x^3 + x^2 - x$$

would be displayed on the screen like this:

```
 4     3    2
x  - 3 x  + x  - x
```

and an integral sign might be displayed like this:

```
/
|       3
|   sin(x ) dx
|
/
```

which represents Maple's attempt to print

$$\int \sin(x^3)dx$$

Derive, which is a PC-based system, can do better than this: it can display a true integral symbol and a range of Greek letters using the PC's own internal character set. However, Derive is virtually alone in providing mathematical symbols. The other computer algebra systems require specialised hardware or front-end software in order to produce true mathematical notation on the screen. As workstations and PCs with graphics displays become more popular, this may change, but at present most computer algebra systems must use "pretty-printing" to display their output.

INTERACTIVE USE

Most people's early experience with computer algebra systems will be gained using the system interactively: the user will run the computer algebra system at the terminal, typing commands and seeing the results (almost) immediately. In interactive mode, Derive is probably the easiest system to learn and use. Most operations in Derive can be carried out via the menus. Indeed, the only time the user regularly needs to type more than a couple of keystrokes is when entering (*authoring*, in Derive jargon) a mathematical expression which is to be the subject of later commands.

The other systems require commands and operator names to be typed in full, which means that the user must remember the names of all the important operators such as differentiation and integration.

INPUT FROM FILES

After a few days or weeks of interactive use, the user will have become familiar enough with the computer algebra system to have developed procedures and definitions for frequently-used operators. A REDUCE user, for example, may have written a procedure to calculate the Taylor series expansion of any function, and will find himself/herself typing the definition at the beginning of every REDUCE session. Fortunately, all of the computer algebra systems provide a means of reading commands and expressions from a file instead of from the terminal. Users can thus store definitions in files, loading them at the start of a session with a single command.

Some calculations performed using computer algebra systems tend to take a very long time to complete, and it can be useful to place the commands for the calculation into a file and run the computer algebra system in the background (or in a batch processing system on some computers), where it takes its instructions from the file rather than from the terminal.

OUTPUT TO FILES

During interactive calculations, the output from the computer algebra system will usually be displayed on the terminal screen. However, it is sometimes useful to be able to save the output by writing it to a file, perhaps for printing at a later date. All of the five computer algebra systems allow the user to send output to a file in the same "pretty-printed" format as they use to show results on the screen.

In addition, most of them can be instructed to save results by writing them to a file in a format that enables the file to be re-read by the computer algebra system during a later session. For most of the computer algebra systems, this simply entails switching off the "pretty-printing" mechanism and writing output using the same format that must be used when typing commands and expressions into the computer algebra system. Some computer algebra systems, however, are able to write results and procedure definitions in a more compact format. This can be re-read much more quickly by the computer algebra system, although it cannot be read or modified directly by the user. REDUCE and Maple use this encoded output technique extensively, and large parts of both systems are supplied in the form of compact binary files instead of the original source code.

FORTRAN AND C OUTPUT

In Chapter 2, it was noted that computer algebra systems cannot perform floating point calculations with the same speed as programs written in conventional languages such as FORTRAN. Many users find that they need to incorporate the results obtained by a computer algebra system into a program written in FORTRAN or C. The computer algebra system may have generated several long and complicated algebraic expressions which must be transcribed into the middle of a FORTRAN or C program. Copying an expression by hand may introduce errors, but fortunately all the computer algebra systems are able to write expressions as if they were FORTRAN or C statements. This includes indenting FORTRAN statements so that they begin in column 7 and adding continuation marks in column 6 when writing an expression that spans two or more lines. Some computer algebra systems can write "optimised" FORTRAN code in which common sub-expressions are extracted and precomputed in separate statements preceding the one which represents the main expression.

REDUCE and MACSYMA incorporate a package called GENTRAN, written by Barbara Gates. This is a very powerful and sophisticated package which can translate REDUCE or MACSYMA commands and program constructs such as loops and if-blocks into the corresponding FORTRAN or C code. It can be used to generate entire FORTRAN or C programs automatically. An example of the use of GENTRAN is given in Chapter 8.

TYPESETTER OUTPUT

Some computer algebra systems are able to take an algebraic expression and generate the instructions needed to typeset the expression using the computer typesetting languages *troff* or TEX. In the case of *troff*, the computer algebra system generates the *eqn* pre-processor instructions, whilst in the case of TEX, the system generates the maths-mode instructions which would be enclosed in '$' symbols. The ability to create TEX or *eqn* code directly from algebraic expressions is a valuable feature, since it relieves the user of much of the difficulty involved in typesetting complicated mathematical formulae when writing papers or books.

GRAPHICS

One of the most important recent developments in the functionality of computer algebra systems has been the provision of a facility for producing high-quality plots of curves and surfaces directly from their algebraic representations. MACSYMA was the first system to offer graphics, but this was limited to Symbolics 3600 workstations. Mathematica, one of the newest computer algebra systems, led the way by providing graphics as a standard part of the system when it was released in 1988. Now, both Mathematica and Maple provide high-quality graphical output in two and three dimensions. They can draw and shade surfaces to give a very realistic impression of perspective and texture. Before the advent of Mathematica, the task of producing such images of surfaces in three dimensions involved the use of specialised graphics packages that were often complicated and difficult to use.

Mathematica can be instructed to produce a picture of a surface with a straightforward command such as

```
Plot3D[Sin[x] Sin[y], {x,-Pi,Pi}, {y,-Pi,Pi},
    Lighting->True]
```

When a more permanent copy of a picture is required, both Mathematica and Maple can generate PostScript files containing the commands needed to reproduce the image on a PostScript-compatible printer.

Derive also incorporates two and three-dimensional graphics. It draws surfaces using the wire-frame method. Using Derive's windowing facility, it is possible to split the PC screen into two or more areas and to display mathematics in one region and graphics in the other. This enables the user to see an expression and view its behaviour at the same time. The illustration on the cover of this book was produced using Derive.

NOTE ON TABLES

The following abbreviations are used in the tables in the next section: **RED** = REDUCE, **MPL** = Maple, **MAC** = MACSYMA, **DER** = Derive, **MTH** = Mathematica.

SUMMARY

Feature	RED	MPL	MAC	DER	MTH
Pretty-printing					
Displays superscripts by pretty-printing	Yes	Yes	Yes	Yes	Yes
Displays integrals by pretty-printing	No	Yes	No	No	No
Displays summations by pretty-printing	No	Yes	Yes	No	No
Displays true integral symbols and Greek letters	No	No	No	Yes	No
Interactive use					
Allows interactive use	Yes	Yes	Yes	Yes	Yes
Input from files					
Read definitions of procedures from files	Yes	Yes	Yes	Yes	Yes
Run entire programs from files	Yes	Yes	Yes	Yes	Yes
Output to files					
Pretty-print to files	Yes	Yes	Yes	Yes	Yes
Write expressions to files and re-read them	Yes	Yes	Yes	Yes	Yes
Write compact binary files	Yes	Yes	Yes	No	No
Save the entire session in a file, to restart later	Yes	Yes	Yes	Yes	Yes

Feature	RED	MPL	MAC	DER	MTH

Fortran and C output

Feature	RED	MPL	MAC	DER	MTH
Generate FORTRAN code	Yes	Yes	Yes	Yes	Yes
Optimise FORTRAN code	No	Yes	No	No	No
Translate system's syntax to FORTRAN	Yes	No	Yes	No	No
Generate C code	Yes	Yes	Yes	No	Yes
Generate Pascal code	No	No	No	Yes	No

Typesetter output

Feature	RED	MPL	MAC	DER	MTH
Generate *eqn* code	No	Yes	No	No	No
Generate TEX code	Yes	Yes	Yes	No	Yes

Graphics

Feature	RED	MPL	MAC	DER	MTH
Graphs of $f(x)$ against x	No	Yes	Yes	Yes	Yes
Multiple graphs	No	Yes	No	Yes	Yes
Wire-frame drawings of surfaces	No	Yes	Yes	Yes	No
Shaded drawings of surfaces	No	Yes	No	No	Yes
Generate PostScript code	No	Yes	No	No	Yes

Chapter 7
Manuals and Help

PRIMER MANUALS

Software packages, especially those as complicated as computer algebra systems, are rarely self-explanatory. All good software should be accompanied by user manuals that are both easy to read and comprehensive in scope.

Several of the systems provide primers which describe their basic features: the syntax of expressions; how to enter commands; how to simplify expressions. Maple and MACSYMA have primer manuals that are separate from the main reference manual, whilst the Mathematica primer is the first chapter of the book *Mathematica: a system for doing mathematics by computer* by Stephen Wolfram. REDUCE has no primer, but Gerhard Rayna's book *REDUCE: Software for algebraic computation* provides an excellent introduction with many examples. Derive is easier to learn to use than the other four systems due to its use of menus, but in any case the user manual is clear and helpful to the beginner.

REFERENCE MANUALS

Experienced users have different requirements to beginners when using software packages. They usually need a specific piece of information as quickly as possible. This might be the syntax of a command or the name of a function. A well-designed reference manual is invaluable to anyone who works frequently with a large software package.

Once again, Maple and MACSYMA possess comprehensive reference manuals, though MACSYMA's reference manual has several indices and it is often difficult to know which index to search in order to locate a particular keyword or function name. REDUCE has a well-organised user manual which describes the system in detail, but it offers only a tantalisingly brief glimpse of the possibilities of extending the REDUCE system by writing in LISP, the language which underlies REDUCE. Mathematica's reference manual is Stephen Wolfram's book cited in the previous section. This is a very readable book and it describes all of the functions and commands that are available, with particular emphasis upon the sophisticated graphics capabilities. Derive's user manual is also thorough and very easy to read.

ON-LINE HELP

Most modern software packages offer on-line help facilities which can provide the user with information whilst the package is in use, complementing or even replacing the reference manual. All of the systems except REDUCE have on-line help facilities. Maple, Mathematica and MACSYMA can provide information on any of their commands and functions. The Maple command

```
help(simplify);
```

displays the description of the `simplify` function given in the *Maple Reference Manual*. The help systems of MACSYMA and Mathematica can also remind the user of the correct name of a command or function if the user can only remember part of the name. The MACSYMA command

```
apropos(int);
```

lists all functions whose names include the string "int". It might result in a list including DEFINT, INTEGRATE and LDEFINT and the user can then request a description of any of these functions.

Derive's help system provides a brief synopsis of the available commands whilst referring the user to the appropriate page of the user manual for more information.

INTERACTIVE TUTORIALS

Some software packages incorporate interactive tutorial lessons which are intended to show the user how the software operates by giving examples and enabling the user to try similar examples within the environment of the packages.

REDUCE, Derive and MACSYMA all provide interactive tutorials demonstrating the use of the system. In both cases, the lessons include pauses where the user can experiment with the commands or concepts that have just been explained, then resume the lesson.

OTHER INFORMATION

The manuals and lessons provided by the software suppliers are not the only source of information about the use of computer algebra systems. Many books, papers and articles have been published by users of these systems to illustrate how they may be used to solve a wide range of problems. The bibliography contains references to many of these, but several deserve special mention.

REDUCE: Software for Algebraic Computation, by Gerhard Rayna, is a comprehensive introduction to REDUCE. It covers all aspects of the system and serves as a useful alternative to the REDUCE User Manual. The current edition of the book describes REDUCE version 3.2. However, more recent versions of REDUCE are available, so there are some discrepancies, notably regarding equation solving. Nevertheless, it is a good reference for newcomers to REDUCE.

Roman Maeder's *Programming in Mathematica* complements the user manual by Stephen Wolfram and explores Mathematica programming techniques in greater detail. Maeder is one of the principal authors of the Mathematica system.

Several books are in preparation which describe the Maple system and the algorithms and programming techniques used by its authors. These include books by Keith Geddes, George Labahn, Gaston Gonnet and Michael Monagan, who are members of the Maple Symbolic Computation Group at the University of Waterloo.

As Derive's potential as a teaching aid is increasingly recognised, books have now begun to appear which are aimed directly at teachers of mathematics. *Exploring Math from Algebra to Calculus with Derive, A Mathematical Assistant* by Jerry Glynn is a very good introduction to Derive, including many examples showing the sequence of key-strokes required to perform various types of calculation. *Calculus and the Derive Program: Experiments with the Computer* by Lawrence Gilligan and James Marquardt is in the form of twenty "laboratory sessions" which use Derive to introduce the student to new concepts in calculus and algebra. The student is guided through each session and must write the "experimental results" on removable pages which can be handed to the teacher for assessment.

Chapter 8

Case Studies

INTRODUCTION

This chapter provides several applications of computer algebra to non-trivial problems. These applications span a variety of disciplines. Different computer algebra systems are exploited in order to demonstrate the different features of each. However, it is important to realise that the choice of a particular computer algebra system was a fairly arbitrary one and that many of the case studies could have been solved by more than one system.

FUNCTION FOR TAYLOR SERIES

One of the features of the REDUCE system is that it has relatively few built-in functions. New users sometimes find this off-putting. However, REDUCE does have a fairly powerful programming language which makes it relatively easy to extend its capabilities.

This case study describes a function for providing REDUCE with the capability for performing a Taylor series expansion. The following procedures provide REDUCE with this capability:

```
PROCEDURE FACT X;
% Returns X factorial
If X < 0 Then Rederr "Illegal argument for factorial."
Else If X <= 1 Then 1
  Else For I := 1:X Product I;

PROCEDURE TAYLOR(FN,X,XVAL,N);
% Computes the Taylor expansion of FN(X) about the point
% X = XVAL. Terms up to order N are sought.
Begin
  Scalar EXPAN;
  % Adjust switches so that each term prints separately
  Off ALLFAC; On DIV,REVPRI;
  DIFF := FN;
  EXPAN := 0;
  For LOOP:=0:N Do Begin
    EXPAN := EXPAN +
```

```
                X**LOOP * SUB(X=XVAL,DIFF) / FACT(LOOP);
      DIFF := DF(DIFF,X,1)
   End;
   Return EXPAN
 End;
```

REDUCE does not provide a procedure for calculating factorials. Fortunately, it is a fairly trivial exercise to provide one using the FOR statement. This is accomplished by the procedure FACT. The procedure TAYLOR returns the Taylor series expansion of a function in a given variable about a specified point. The final argument specifies the desired number of terms. Thus

```
TAYLOR(sin(x),x,0,10)
```

returns

$$x - \frac{1}{6}x^3 + \frac{1}{120}x^5 - \frac{1}{5040}x^7 + \frac{1}{362880}x^9$$

The procedure TAYLOR contains a loop which builds up the expansion in the variable EXPAN. It is worth noting that the procedure stores the previous derivative in the variable DIFF. This provides important advantages in performance, as repeatedly calculating higher derivatives of a complicated expression can be very time consuming. Indeed, the same technique could have been used in calculating the factorial term in the denominator. However, the potential saving here is much smaller.

GENERATING FORTRAN CODE

Computer algebra systems provide scientists and engineers with the means to find exact algebraic solutions to a wide range of problems that might otherwise be soluble only by using approximate numerical methods. However, it should not be thought that algebraic methods and numerical methods are mutually exclusive. This is illustrated by the following case study in which REDUCE is used to create FORTRAN subroutines automatically with the aid of the GENTRAN package. GENTRAN is supplied as part of the REDUCE system.

The first example involves numerical interpolation of a function whose values are given at several equally-spaced abscissae. Suppose that the function is known at an even number of points, say $2N$. The abscissae of these points may be denoted by

$$X_{-N+1}, \cdots, X_{-1}, X_0, X_1, X_2, \cdots, X_N$$

where

$$X_j = X_0 + jh$$

and h is the interval between pairs of abscissae. The value of the function at the jth point, $f(X_j)$, is denoted by the shorthand notation f_j. That is to say,

$$f_j = f(X_j) = f(X_0 + jh)$$

Lagrange's interpolation formula enables the value of the function at some point

$$X_0 + ph \qquad (0 \leq p \leq 1)$$

to be approximated by the expression

$$f(X_0 + ph) \simeq \sum_{k=-N+1}^{N} \mathcal{L}_k^{2N}(p) f_k$$

where $\mathcal{L}_k^{2N}(p)$ is a polynomial of degree $2N - 1$ in p known as a Lagrange interpolation polynomial. It can be written in the following form:

$$\mathcal{L}_k^{2N}(p) = (-1)^{N+k} \binom{2N}{N+k} \frac{l_N(p)}{p-k}$$

where

$$l_N(p) = \frac{1}{N!} \prod_{j=-N+1}^{N} (p-j)$$

It would be time-consuming, and possibly error-prone, to calculate these polynomials by hand for values of N greater than 2 or 3.

However, computer algebra systems make light work of such calculations. Here is a REDUCE program which uses GENTRAN to create a FORTRAN subroutine to calculate the Lagrange interpolation polynomials. The subroutine is generated for a fixed value of N which must be specified by the user before running the REDUCE program. In this example, $N = 3$ and hence the polynomials are generated for a six-point Lagrange interpolation scheme.

```
% GENTRAN program to generate Lagrange interpolation
% polynomials

% Load the GENTRAN package
load gentran$

% Prevent REDUCE from echoing input
off echo$

% Retain polynomials in factorised form
on factor$

% Define the factorial operator and the binomial
% coefficients ("n CHOOSE m")
operator fac$
for all n such that fixp n and n>0 let fac n =
  for j:=1:n product j$
let fac 0 = 1$

operator choose$
for all m,n such that fixp m and fixp n and m>=0 and n>=0
  and m>=n let choose(m,n) = fac(m)/(fac(n)*fac(m-n))$

% This is the number of data points - the user selects the
% value of this variable
n := 6$

m:=n/2$

% Open the FORTRAN file
gentranout "lpoly.for"$
```

```
% Specify the maximum line length
fortlinelen!* := 45$

% Generate the subroutine statement
gentran literal tab!*,"SUBROUTINE INTERP(V,X,H)",cr!*$

% Generate the type declaration statements
gentran literal tab!*,"REAL V(",eval(-m+1),":",
  eval(m),"),X,H,L,P",cr!*$

gentran p:=x/h$
l:=(for j:=-m+1:m product (p-j))/(fac n)$
for j:=-m+1:m do % Generate each polnomial
<<   q := (-1)**(N+j) * choose(n-1,m-1+j) * l/(p-j)$
     gentran v(eval(j)) := eval q>>$

% Generate the RETURN and END statements
gentran literal tab!*,"RETURN",cr!*,tab!*,"END",cr!*$

% Close the FORTRAN file
gentranshut "lpoly.for"$

quit$
```

The resulting output is a complete FORTRAN subroutine which can be compiled and incorporated into a larger program:

```
SUBROUTINE INTERP(V,X,H)
REAL V(-2:3),X,H,L,P
P=X/H
V(-2)=(P+1.000000E0)*(P-1.000000E0)*(P-
. 2.000000E0)*(P-3.000000E0)*P/
. 7.200000E2
V(-1)=-((P+2.000000E0)*(P-1.000000E0)*(
. P-2.000000E0)*(P-3.000000E0)*P)/
. 1.440000E2
V(0)=(P+2.000000E0)*(P+1.000000E0)*(P-
. 1.000000E0)*(P-2.000000E0)*(P-
. 3.000000E0)/7.200000E1
V(1)=-((P+2.000000E0)*(P+1.000000E0)*(P
. -2.000000E0)*(P-3.000000E0)*P)/
. 7.200000E1
```

```
V(2)=(P+2.000000E0)*(P+1.000000E0)*(P-
. 1.000000E0)*(P-3.000000E0)*P/
. 1.440000E2
V(3)=-((P+2.000000E0)*(P+1.000000E0)*(P
. -1.000000E0)*(P-2.000000E0)*P)/
. 7.200000E2
RETURN
END
```

UNITS

One of the disadvantages of the use of pocket calculators has been that the units of physical quantities have tended to become divorced from their magnitude.

For example, a student who needs to derive a value for the Bohr magneton:

$$a_0 = \frac{h^2}{\pi \mu_0 c^2 m_e e^2}$$

would substitute relevant numbers into this formula and subsequently attach a unit of length to the answer.

A computer algebra system enables the units to remain associated with the quantity. This can be demonstrated by performing this calculation with REDUCE. The following program performs the calculation and retains the association between the quantity and its units:

```
% Prints out the Bohr radius;
ON BIGFLOAT;
ON NUMVAL;
% Define various physical constants;
me := 9.109390 * (10**-31) * kg$
charge := 1.602177 * (10**-19) * Coul$
mu0 := 4 * PI * (10**-7) * N * (s/Coul)**2$
c := 2.997925 * (10**8) * m/s$
h := 6.626075 * (10**-34) * J * s$
% Define various units;
J := m * N$
```

```
N := kg * m / s**2$
bohrrad := h**2 / ( PI * muO * c**2 * me * charge**2);
;END;
```

It returns the answer:

$$a_0 = 5.2917810^{-11} \ m$$

The fact that the units are correct makes it unlikely that a quantity has been omitted or has been included in the numerator rather than the denominator (or vice versa).

Associating the units with a quantity also makes it simple to convert from one system of units to another. For example, the previous answer can be converted to feet or Angstroms by including a definition for the metre.

DIMENSIONAL ANALYSIS

Dimensional analysis is a technique which can be used to predict product relationships. It can only reveal the complete form of a relationship if less than five independent quantities are used.

The first application of dimensional analysis is to the flow of liquid through a circular tube. The rate of flow V through a circular tube depends upon the pressure difference P between the ends of the tube, the radius of the tube r and the viscosity of the liquid η. This can be written in the form:

$$V = k\eta^X r^Y P^Z$$

where k is a dimensionless constant. Dimensional analysis is used to determine the powers X, Y and Z. The following Maple program can be used to accomplish this:

```
# Maple program for performing dimensional analysis

# Define equation for rate of flow through a pipe
floweqn := VOLUME/T = K * VISCOSITY**X *
  RADIUS**Y * PRESSUREGRAD**Z;
```

```
# Define quantities used in the flow equation
VOLUME := L**3;
VISCOSITY := FORCE/(AREA * VEL * 1/L);
FORCE := M*ACCN;
ACCN := L/T**2;
AREA := L**2;
VEL := L/T;
RADIUS := L;
PRESSUREGRAD := PRESSURE/L;
PRESSURE := FORCE/AREA;

# Take logs of both sides and expand logs of powers
map(log,floweqn);
simplify(");
floweqn:=subs(ln(K) = 0,");

# Obtain an equation for M by eliminating L and T
meqn := subs(L=1,T=1,floweqn);

# Obtain an equation for L by eliminating M and T
leqn := subs(M=1,T=1,floweqn);

# Obtain an equation for T by eliminating M and L
teqn := subs(M=1,L=1,floweqn);

# Solve the three equations obtained above
solve({meqn,leqn,teqn},{X,Y,Z});
```

The program begins by introducing the flow equation. All quantities in the flow equation are then defined in terms of the three fundamental quantities mass, length and time denoted by M, L and T. Upon substitution into the flow equation, the intermediate result

$$L^3 T^{-1} = k M^{X+Z} L^{-X+Y-2Z} T^{-X-2Z}$$

is obtained. In a conventional pen and paper solution, the technique of "inspection" would be used to obtain three simultaneous equations for X, Y and Z. With Maple (or indeed any computer algebra system) it is necessary to work a little harder to obtain the equations. The stages in extracting the simultaneous equations are as follows:

- Take the log of both sides of the equations, using the map function.
- Apply the simplify function which turns expressions of the form

$$\log(a^b)$$

into

$$b \log a$$

- Remove the log of the dimensionless constant, k.
- Evaluate the resulting expression setting $L = T = 1$ to obtain an equation for the powers of M.
- Repeat the previous calculation setting $M = T = 1$ to obtain an equation for the powers of L.
- Repeat the previous calculation setting $M = L = 1$ to obtain an equation for the powers of T.

Finally, the three simultaneous equations are solved for the three unknowns. This produces the result

$$X = -1,$$
$$Y = 4,$$
$$Z = 1.$$

Thus the original flow equation must be of the form

$$V = \frac{kr^4 P}{\eta}.$$

This result is consistent with Poiseuille's formula which states that:

$$V = \frac{\pi r^4 P}{8\eta}.$$

Whilst dimensional analysis alone can only be applied to situations in which there are not more than four independent quantities, it still retains some value when there are more than four independent quantities. The next example of dimensional analysis has five independent quantities. However, dimensional analysis combined with the application of physical laws enables a solution to be found.

The force F on a sphere moving through a liquid depends on the velocity of the sphere v, the viscosity of the liquid η, the density of the liquid ρ and the radius of the sphere r. This can be expressed as:

$$F = kv^A\eta^B\rho^C r^D$$

The following Maple program can be used to find solutions for A, B, C and D.

```
# Maple program for performing dimensional analysis

# Define equation for force on a body moving through a
# viscous fluid
retardeqn := FORCE = K * VELOCITY**A * VISCOSITY**B *
                DENSITY**C * RADIUS**D;

# Define substitutions to be used
FORCE := M*ACCN;
ACCN := L/T/T;
AREA := L**2;
VOLUME := L**3;
substit := {
  VISCOSITY = FORCE/(AREA * L/T * 1/L),
  VELOCITY = L/T,
  DENSITY = M/VOLUME,
  RADIUS = L
};
subs(substit,retardeqn);

# Take logs of both sides and expand logs of powers
map(log,");
simplify(");
logretardeqn:=subs(ln(K) = 0,");

# Obtain an equation for M by eliminating L and T
meqn := subs(L=1,T=1,logretardeqn);

# Obtain an equation for L by eliminating M and T
```

```
leqn := subs(M=1,T=1,logretardeqn);

# Obtain an equation for T by eliminating M and L
teqn := subs(M=1,L=1,logretardeqn);

# Solve the three equations obtained above in terms of A
solve({meqn,leqn,teqn},{B,C,D});

# Put these solutions back into the original equation
subs(op("),retardeqn);

# Consider stream line flow (A=1)
subs(A=1,");

# Consider turbulent flow (A=2)
subs(A=2,"");
```

The program has many similarities with the previous example and so it is only necessary to highlight the differences.

- It is important in this case to retain the original equation for subsequent use, so only temporary substitutions of the velocity, viscosity, density and radius are made with a subs command.
- Solutions for B, C and D are obtained in terms of A.
- These solutions are then substituted back into the original equation.

Fluid flow is often considered to be either laminar or turbulent. Laminar flow occurs at low velocities and in this situation the velocity is directly proportional to the force. Turbulent flow occurs at high velocities and in this situation the square of the velocity is directly proportional to the force. The program concludes by taking each of these cases in turn. For laminar flow it is found that

$$V = kv\eta r$$

The independence of the force on the density of the liquid is in good agreement with experiment. For turbulent flow the following result is obtained:

$$F = kv^2 \rho r^2$$

This time the independence of the force on viscosity of the liquid is once again in good agreement with experiment results.

SOLVING ODES

For applications to mechanics there is the alternative of using computer algebra systems to complement theoretical and numerical approaches.

The simplest model of free fall of a body in the Earth's gravitational field leads to the one dimensional differential equation

$$\frac{d^2 x}{dt^2} = g$$

where x is the distance, t is the time and g is the constant gravitational acceleration. Introducing the velocity

$$v = \frac{dx}{dt}$$

sets up the simple iterative model

$$x = x + vh$$
$$v = v + gh$$
$$t = t + h$$

where h is an arbitrary small step width. The initial assignments are

$$x = 0,$$
$$v = 0,$$
$$t = 0.$$

Already an error has been introduced by calculating x before v in the sequence. This causes values of the distance and velocity to be lower than their exact values. However, changing the order of v and x will not affect v but will produce slightly higher values for x so the average of the two x values should be a good estimate of the true value. Even so, in either case the error should decrease as the number of steps is increased, i.e., h is made smaller. After n iterations the value of t will be nh so x and v can be found as functions of t by substituting $h = t/n$ in the final iterates.

The following REDUCE program finds the upper and lower bounds and the average for x. The switches in the first line are set to produce a better display.

```
% Set display switches
off allfac; on div; on revpri;
% Initial assignments
xl:=xu:=0;vl:=0;
% Assign the number of steps and evaluate loop
n:=10;
for i:=1:n do <<xl:=xl+v*h;v:=v+g*h;xu:=xu+v*h>>;
% Substitute for time and find the average value of the
% distance
xl:=sub(h=T/n,xl);
xu:=sub(h=T/n,xu);
v:=sub(h=T/n,v);
x:=(xl+xu)/2;
;end;
```

This gives a lower value of

$$x_l = \frac{9gt^2}{20}$$

and an upper value of

$$x_u = \frac{11gt^2}{20}$$

with an average value of

$$x = \frac{gt^2}{2}$$

while

$$v = gt$$

From this elementary example it is obvious that both the upper and lower values of x are converging on the standard formula

$$x = \frac{gt^2}{2}$$

and the average is exact. The velocity is exact. Increasing n will only demonstrate the convergence of the upper and lower bounds on x. With n assigned to 20,

$$x_l = \frac{19gt^2}{40}$$

$$x_u = \frac{21gt^2}{40}$$

The extensions to similar problems offer many possibilities: different initial conditions, projectiles in two and three dimensions, air resistance and central orbit problems.

Now consider the introduction of air resistance proportional to the velocity. In the example above, only the loop needs changing to

```
for i:=1:n do <<xl:=xl+v*h;v:=v+(g-k*v)*h;xu:=xu+v*h>>;
```

to include the resistance term. Running the program with $n = 5$ gives

$$x = \frac{1}{2}gt^2 - \frac{3}{25}gkt^3 + \frac{2}{125}gk^2t^4 - \frac{7}{6250}gk^3t^5 + \frac{1}{31250}gk^4t^6$$

and

$$v = gt - \frac{2}{5}gkt^2 + \frac{2}{25}gk^2t^3 - \frac{1}{125}gk^3t^4 + \frac{1}{3125}gk^4t^5.$$

The answers are appreciably more complicated and the existence of analytical solutions is more doubtful. Experiments can be tried, looking at v, it is an alternating series, only the first term can be considered to have converged so maybe higher powers of n can be tried. This will cause truncation error problems, but

REDUCE's LET command is very useful in this regard. Putting
the command

```
let k**4 = 0;
```

in the second line of the modified program truncates all series at
the third power of k running again with $n = 100$ gives

$$x = \frac{1}{2}gt^2 - \frac{6567}{40000}gkt^3 + \frac{160083}{4000000}gk^2t^4 - \frac{30899253}{4000000000}gk^3t^5$$

and

$$v = gt - \frac{99}{200}gkt^2 + \frac{1617}{10000}gk^2t^3 - \frac{156849}{4000000}gk^3t^4$$

With $n = 500$ the results are

$$x = \frac{1}{2}gt^2 - \frac{166167}{1000000}gkt^3 + \frac{20667083}{500000000}gk^2t^4 - \frac{20522496253}{2500000000000}gk^3t^5$$

and

$$v = gt - \frac{499}{1000}gkt^2 + \frac{41417}{250000}gk^2t^3 - \frac{20584249}{500000000}gk^3t^4$$

For v the coefficient of gkt^2 is approximately 0.499 and the coefficient of gk^2t^3 is 0.165668. These are close to $1/2$ and $1/6$
suggesting that v is related to a power series involving the exponential function. Increasing n would improve the convergence,
but this is tantamount to integrating so an entirely new way of
producing the results uses REDUCE's integration facility.

First a procedure for definite integration, then the obvious extension of the finite time steps to the continuous integration

```
% Definite integration procedure
procedure defint(f,x,ll,ul);
begin scalar prim;
      prim:=int(f,x);
      return sub(x=ul,prim)-sub(x=ll,prim);
end;
% Initial assignments
xi:=0;v:=0;vi:=0;
n:=5;
```

```
% Evaluate the assigned number of repeated integrations
for i:=1:n do <<x:=xi+defint(v,t,0,t);
                 v:=vi+defint((g-k*v),t,0,t)>>;
x;v;
;end;
```

The program generates the following expressions for x and v:

$$x = \frac{1}{2}gt^2 - \frac{1}{6}gkt^3 + \frac{1}{24}gk^2t^4 - \frac{1}{120}gk^3t^5$$

$$v = gt - \frac{1}{2}gkt^2 + \frac{1}{6}gk^2t^3 - \frac{1}{24}gk^3t^4$$

The exponential form of both variables looks obvious. The following program checks the value of v. A similar program can be written to check the value of x.

```
% Ad hoc factorial procedure
procedure fac(n); for i:=1:n product i;
% Generate the power series for exp(-kt) to check the
% velocity
checkv:=for i:=1:n sum(g/k*(-k*t)**i/fac(i));
;end;
```

Starting with discrete steps and generalising to integration is possible in many other situations. For example, the standard simple harmonic motion differential equation

$$\frac{d^2x}{dt^2} + x = 0$$

can be represented as two coupled first-order differential equations:

$$v = \frac{dx}{dt},$$

$$\frac{dv}{dt} = -x.$$

If the initial conditions are $x = 0$, $v = 1$ at $t = 0$, a discrete program, using the above display setting, which underestimates the value of x is

```
% Set the display switches
on revpri; on div; off allfac;
% Assign the initial values
xl:=0;v:=1;
n:=5;
% Evaluate the assigned number of integrations
for i:=1:n do <<xl:=xl+v*h;v:=v-xl*h>>;
% Substitute for time
xl:=sub(h=T/n,xl);
vl:=sub(h=T/n,vl);
;end;
```

The output is

$$x_l = t - \frac{4}{25}t^3 + \frac{21}{3125}t^5 - \frac{8}{78125}t^7 + \frac{1}{1953125}t^9$$

$$v_l = 1 - \frac{3}{5}t^2 + \frac{7}{125}t^4 - \frac{28}{15625}t^6 + \frac{9}{390625}t^8 - \frac{1}{9765625}t^{10}$$

The answers suggest the trigonometric power series, but the co-efficients have not converged. Increasing n is a possibility, but the corresponding program using integration,

```
xi:=0;v:=vi:=1;
for i:=1:n do <<x:=xi+defint(v,t,0,t);
                v:=vi-defint(x,t,0,t)>>;
x;v;
;end;
```

produces

$$x = t - \frac{1}{6}t^3 + \frac{1}{120}t^5 - \frac{1}{5040}t^7 + \frac{1}{362880}t^9$$

and

$$v = 1 - \frac{1}{2}t^2 + \frac{1}{24}t^4 - \frac{1}{720}t^6 + \frac{1}{40320}t^8 - \frac{1}{362880}t^{10}.$$

which are immediately recognisable as the first few terms in the power series for sine and cosine respectively.

This ability to recast differential equations as iterative integral equations has many applications. Perturbation methods have become more tractable with computer algebra systems, as the work of Fitch (1985) and Rand (1984) demonstrate as well as many other examples scattered throughout the applications literature.

Other techniques for solving differential equations that require routine but tedious integration such as the Galerkin method are also made faster, easier and less error prone as Mills has demonstrated. Also two of the authors have used REDUCE to produce FORTRAN subroutines to calculate the Jacobian matrices of derivatives needed in Gear's method of numerically integrating differential equations. This case study has illustrated the connection between discrete and continuous methods of solving differential equations and suggested that selected use of computer algebra systems may enhance or speed up numerical methods.

AC CIRCUITS

Computer algebra systems are ideally suited to solving the sets of linear simultaneous equations which arise from analysing AC circuits.

The Wien bridge is a typical AC bridge in which two adjacent arms contain resistors, whilst the remaining arms contain a series and parallel combination of impedances.

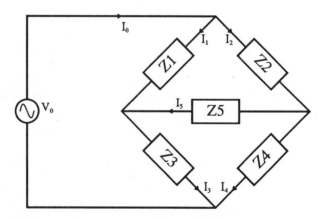

Figure 1: The Wien bridge

If Kirchoff's laws are applied to the AC circuit shown in figure

1, then the following six independent equations are satisfied:

$$I_0 - I_1 - I_2 = 0$$
$$I_0 - I_3 - I_4 = 0$$
$$I_1 - I_3 + I_5 = 0$$
$$V_0 - I_1 Z_1 - I_3 Z_3 = 0$$
$$V_0 - I_2 Z_2 - I_4 Z_4 = 0$$
$$I_1 Z_1 - I_2 Z_2 - I_5 Z_5 = 0$$

The following Maple program can be used to find the values of the resistor R_2 and frequency ω which balance the Wien bridge.

```
# Solves 6 equations associated with the Wien bridge

# Procedure for summing Impedances in Series
SeriesImp := proc()
  local loop;
  sum('args[loop]','loop'=1..nargs)
end;

# Procedure for summing Impedances in Parallel
ParallelImp := proc()
  local loop;
  1/sum(1/'args[loop]','loop'=1..nargs)
end;

# Returns real part of an expression
RealPart := proc(expr)
  evalc(subs(j=I,map(Re,expr)))
end;

# Returns Imaginary part of an expression
ImagPart := proc(expr)
  evalc(subs(j=I,map(Im,expr)))
end;

# Provide 6 equations which specify the currents
# and voltages
E1 := I0 - I1 - I2;
E2 := I0 - I3 - I4;
```

```
E3 := I1 - I3 + I5;
E4 := V0 - I1*Z1 - I3*Z3;
E5 := V0 - I2*Z2 - I4*Z4;
E6 := I1*Z1 - I2*Z2 - I5*Z5;
```

```
# Find the current through the detector
currents := solve({E1,E2,E3,E4,E5,E6},
                  {I0,I1,I2,I3,I4,I5});
# Isolate the detector current
readlib(select):
select(has,currents,I5);
op(1,");
op(2,");

# Find the value of Z1 which gives the balance
# (no current)
(Z1 = solve(",Z1));

# Define the 4 impedances for a Wien bridge and
# substitute them
subs(Z1=R1,Z2=R2,Z3=ParallelImp(R3,-j/Omega/C3),
    Z4=SeriesImp(R4,-j/Omega/C4),");

# Determine the values of R2 and Omega which satisfy
# the real and imaginary parts.
solve({RealPart("),ImagPart(")},{Omega,R2});
```

The Maple program starts by defining four simple functions. SeriesImp and ParallelImp are generalised to deal with combinations of series and parallel impedances. They exploit the fact that a set of series impedances satisfies

$$Z_{total} = Z_1 + Z_2 + Z_3 + \cdots$$

whilst a series of parallel impedances satisfies:

$$\frac{1}{Z_{total}} = \frac{1}{Z_1} + \frac{1}{Z_2} + \frac{1}{Z_3} + \cdots$$

Electrical engineers normally use the symbol j rather than i to refer to $\sqrt{-1}$. This program adopts the same convention. As a result, two procedures RealPart and ImagPart are defined.

These replace j by i immediately before using the relevant Maple commands for extracting the real and imaginary parts of an expression.

The program then defines the six equations given previously. Maple's `solve` command is used to obtain solutions for the six currents. The current flowing through the detector I_5 is of interest here. It can be isolated from the other solutions by using the `select` and `op` commands. The `solve` is again applied to the resulting equation for the case where the detector current is zero. The `solve` command is used to obtain an expression for Z_1 in terms of the other impedances. However, one of the impedances in a different arm could equally well have been chosen.

Although this program is intended to solve the Wien bridge circuit, all the analysis thus far is applicable to any AC bridge. This has the advantage that minimal changes would be needed to adapt it for solving a different combination of impedances. The `subs` command makes the program specific to the Wien bridge by substituting a specified combination of resistors and capacitors in place of the four impedances. Finally `solve` is again used to obtain solutions for the real and imaginary parts of the result. The values of R_2 and frequency ω which balance the Wien bridge are as follows:

$$R_2 = \frac{R_1(C_4R_4 + C_3R_3)}{C_4R_3}$$

$$\omega = \frac{1}{\sqrt{C_3R_3C_4R_4}}$$

The authors gratefully acknowledge their debt to Dr. Francis Wright of Queen Mary & Westfield College London whose ideas have been used in this example.

TENSOR CALCULUS

The manipulation of tensors is a complicated and lengthy process. In the four dimensional manifold of general relativity, even a second-rank tensor has sixteen components, and a general fourth-rank tensor has 256 components. Naturally, computer algebra systems can play a valuable role in relieving the user of the need to carry out countless additions, multiplications and differentiations when working with tensors. Some packages, such as SHEEP and STENSOR, are designed principally with tensor algebra and calculus in mind. Most of the general-purpose computer algebra systems do not have built-in functions for tensor manipulation, but it is relatively easy to define them. This example shows a Maple function which takes a metric tensor as its input and yields the differential equations for the geodesics in the space described by the metric. It applies one of the two standard forms of the geodesic equations:

$$\frac{d^2 x^i}{ds^2} = -\left\{ \begin{matrix} i \\ l \ k \end{matrix} \right\} \frac{dx^l}{ds} \frac{dx^k}{ds}$$

$$\frac{d}{ds}\left(g_{ij} \frac{dx^j}{ds} \right) = \frac{1}{2} \frac{\partial g_{kl}}{\partial x^i} \frac{dx^l}{ds} \frac{dx^k}{ds}$$

The second form is the most convenient if the metric is diagonal, whilst the first form gives the second derivative of each coordinate directly. The procedure chooses the most appropriate form of the geodesic equations depending upon the properties of the metric.

The procedure illustrates several features which commonly appear in procedures written for computer algebra systems, so it will be analysed in detail.

All procedures which are written for other people to use should begin with a description of the purpose of the procedure and the way in which it should be invoked:

```
# Procedure to calculate the geodesic equations
# from a given covariant metric
#
# Usage: geodesics(metric,coordinates,pathvar)
#
#   'metric'   is a square array of dimension 1..N,1..N
#              containing the covariant components of the
#              metric tensor
#
```

```
#   'coordinates'
#                is a vector of dimension 1..N containing the
#                names of the coordinate variables
#
#   'pathvar' is the path variable
#
```

Now the procedure name and its formal parameters are declared, together with the variable names that will be used within the procedure and which are to be regarded as local.

```
geodesics := proc(g,x,s)
   local guu,chr1,chr2,geod,n,i,j,k,l,lhs,rhs,coordset;
```

The first task when writing a robust procedure is to check that the user has supplied a valid set of arguments. Maple's **type** and op functions may be used to determine the type of each argument and, if it is an array, its dimensions.

```
# First, check that the first argument is an array
if not type(g,array) then
  ERROR('first argument not an array') fi:
# Check its dimensions
j := [op(2,op(g))]:
# This should have two parts, both of the form 0..n
if nops(j)<>2 then
  ERROR('first argument has wrong dimension') fi:
j := convert(j,set):
if nops(j)<>1 then
  ERROR('first argument is not a square array') fi:
j := op(1,j):
if op(1,j)<>1 then
  ERROR('lower array bound of metric is not one') fi:
n := op(2,j):
if not (type(n,integer) and n>1) then
  ERROR('upper array bound of metric is invalid') fi:

# Now check that the second argument is a vector
if not type(x,array) then
  ERROR('second argument is not an array') fi:

j := [op(2,op(x))]:
```

```
if nops(j)<>1 then
  ERROR('second argument is not a vector') fi:
j := op(1,j):
if op(1,j)<>1 then
  ERROR('lower bound of second argument is not one') fi:
if op(2,j)<>n then
  ERROR('upper bound of second argument is invalid') fi:

# We're now sure that all arguments are valid
```

It is now necessary to determine whether the metric tensor has
been specified as a diagonal array. The indexfunc function in
the linear algebra package linalg extracts this information.

```
if linalg[indexfunc](g)='diagonal' then
```

If the metric is diagonal, the geodesic equations may be calculated using the second form given above.

```
# It's diagonal, so we use the form of the geodesic
# equations which do not involve the Christoffel symbols
#
geod := array(1..n);
coordset := {coords[i]$i=1..n}:
for i from 1 to n do
  if has(g[i,i],coordset) then
    # This component of the metric contains one or
    # more of the coordinates
    lhs := Diff(g[i,i]*Diff(x[i],s),s)
  else
    # This component of the metric is independent
    # of the coordinates
    lhs := g[i,i]*Diff(x[i],s,s)
  fi:
  rhs := 0:
  for j from 1 to n do
    for k from 1 to n do
      rhs := rhs +
              diff(g[j,k],x[i])*Diff(x[j],s)*Diff(x[k],s)
    od
  od:
  geod[i] := lhs = 1/2*rhs
od
```

Otherwise, if the user has not declared the metric to be diagonal, then the first form of the geodesic equations is used.

```
else
  # It's not diagonal, so we use the form of the geodesic
  # equations which involve the Christoffel symbols
  #
  # Calculate the contravariant metric
  guu := linalg[inverse](g):

  # Calculate the Christoffel symbols of the first kind
  chr1 := array(1..n, 1..n, 1..n):
  for i from 1 to n do
    for l from 1 to n do
      for k from 1 to n do
        chr1[i,k,l] := 1/2*(diff(g[i,l],x[k])
                            +diff(g[k,l],x[i])
                            -diff(g[i,k],x[l]))
      od
    od
  od:

  # Calculate the Christoffel symbols of the second kind
  chr2 := array(1..n, 1..n, 1..n):
  for j from 1 to n do
    for i from 1 to n do
      for k from 1 to n do
        chr2[j,i,k] := sum('guu[j,l]*chr1[i,k,l]',
                           'l'=1..n)
      od
    od
  od:

  # Calculate the geodesics
  geod := array(1..n):
  for i from 1 to n do
    j := Diff(x[i],s,s):
    for l from 1 to n do
      for k from 1 to n do
        j := j + chr2[i,l,k]*Diff(x[l],s)*Diff(x[k],s)
      od
    od:
```

```
   geod[i] := j = 0
   od
fi: # end if (metric is not diagonal)
```

The final task is to return the array containing the geodesic equations, in whatever form they have been calculated.

```
RETURN(op(geod))
end:
```

The procedure uses a dummy differentiation function *Diff* when calculating the derivatives of the coordinates x^i with respect to the path variable s since Maple does not allow the user to declare that the x^i depend upon s in a convenient way. A term such as

```
diff(x[j],s)
```

would evaluate to zero, so the procedure represents it as

```
Diff(x[j],s)
```

As an illustration of the use of this procedure, the equations for geodesics on the surface of the unit sphere may be calculated using the following commands:

```
metric:=array(diagonal, 1..2, 1..2, [(1,1)=1,
   (2,2)=sin(phi)**2]):
coordinates:=array(1..2,[phi,theta]):
geodesics(metric, coordinates, 's');
```

The procedure uses the second form of the geodesic equations, yielding

$$\frac{d^2\theta}{ds^2} = \sin\theta\cos\theta\left(\frac{d\phi}{ds}\right)^2$$

$$\frac{d}{ds}\left(\sin^2\theta\frac{d\phi}{ds}\right) = 0$$

If the metric had not been declared as diagonal, then the procedure would use the first form of the geodesic equations, to yield

$$\frac{d^2\theta}{ds^2} = \sin\theta\cos\theta\left(\frac{d\phi}{ds}\right)^2$$

$$\frac{d^2\phi}{ds^2} = -2\cot\theta\frac{d\theta}{ds}\frac{d\phi}{ds}$$

Inspection shows the second equation of this pair to be the same
as the second equation of the previous pair after expansion of the
derivative.

Chapter 9
Bibliography

This chapter provides a bibliography of books and papers on computer algebra. Where possible, each reference contains a brief summary. The bibliography is not intended to be exhaustive. In particular, many papers which deal with low-level algorithms and specialised applications have been omitted.

There are sections for each of the major systems, together with sections on introductory papers and teaching use. Finally, there is information on newsletters and electronic discussion groups.

Some of the information presented here is based on a similar, but shorter bibliography which was first generated by Jane Bryan-Jones and Francis Wright.

INTRODUCTORY WORKS

MacCallum, M.A.H., *Computer algebra – tomorrow's calculator?*, **New Scientist**, **112**, 52-55 (23 October 1986)

An article which considers why computer algebra has had such a long infancy. An example of the use of the SHEEP system in general relativity is provided.

MacCallum, M.A.H., *Pocket calculus*, **Physics World**, **2(6)**, 27-29 (June 1989)

This provides an introduction to computer algebra. It includes several examples of REDUCE.

Pavelle, R., Rothstein, M. and Fitch, J., *Computer algebra*, **Scientific American**, **245**, 136-152 (December 1981)

A very readable introduction to computer algebra which includes a brief history.

DERIVE

Gilligan, L.G. and Marquardt, J.F., *CALCULUS and the DERIVE Program: Experiments with the Computer* (Gilmar Publishing Company, 1990)

As the title suggests, this book can be used to teach calculus with Derive. It is essentially a work book with pull out pages and spaces for solutions. There are 20 lessons covering differentiation, integration and series.

Glynn, J., *Exploring Math from Algebra to Calculus with Derive, A Mathematical Assistant* (MathWare, 1989)

This book provides a basic introduction to the facilities in Derive. Very few books are suitable for teachers who wish to introduce computer algebra to students in their early teens. Many of the chapters in this book are ideal for this age group.

Harper, D., *Maths on the Menu*, **Physics World**, **3(2)**, 43 (February 1990)

A review of Derive.

Wooff, C.D. and Hodgkinson, D.E., *muMATH: A Microcomputer Algebra System* (Academic Press, 1987)

This is an introduction to the muMATH: the first computer algebra system to become available on micros. It includes many examples of the use of muMATH, including case studies from the authors' own experience.

MACSYMA

Drinkard, R.D. and Sulinski, N.K., *MACSYMA: A Program for Computer Algebra Manipulations* (Naval Underwater Systems Center, Newport, Rhode Island, NUSC Technical Document 6401, 1981)

A ninety-eight page document with a wide variety of demonstrations intended to highlight the basic features of the MACSYMA program. A good introduction to MACSYMA.

Hussain, M. A. and Noble, B., *Applications of MACSYMA to calculations in applied mathematics* (GEC Report 83CRD054, 1979)

A paper which contains a collection of ten different advanced problems in applied mathematics solved using MACSYMA. There are extensive references to further work.

Pavelle, R. and Wang, P. S., *MACSYMA from F to G*, **Journal of Symbolic Computation**, 1, 69-100 (1985)

A descriptive non-technical tutorial on MACSYMA.

Rand, R.H., *Computer Algebra in Applied Mathematics: an Introduction to MACSYMA* (Pitman Research Notes in Mathematics **94**, 1984)

Contains several different applications of MACSYMA with detailed examples of perturbation methods.

Rand, R.H. and Armbruster, D., *Perturbation Methods, Bifurcation Theory and Computer Algebra* (Springer Applied Mathematical Science Series **65**, 1987)

Sloane, N.J.A., *My friend MACSYMA*, **Notices of the American Mathematical Society**, **32**, 40-43 (1985)

Stoutemyer, D.R., *Dimensional Analysis, Using Computer Symbolic Mathematics*, **Journal of Computational Physics**, **24**, 141-149 (1977)

A paper which provides a program in MACSYMA for automatic dimensional analysis.

Symbolics, Inc., *Bibliography of Papers Referencing MACSYMA*, (Symbolics Inc.)

A very detailed list of papers using MACSYMA produced and updated by the MACSYMA group at Symbolics, Inc.

MAPLE

Char, B.W., Fee, G.J., Geddes, K.O., Gonnet, G.H. and Monagan M.B., *A Tutorial Introduction to Maple*, **Journal of Symbolic Computation**, **2**, 179-200 (1986)

A description of the Maple computer algebra system by some members of the Maple project.

Geddes, K.O., *On the Design and Performance of the Maple System*, in **Proceedings of the 1984 MACSYMA Users' Conference**, 199, edited by Ellen V. Golden (1984)

A description of the design of Maple which demonstrates how efficient Maple can be.

Harper, D., *Maple User Guide* (Manchester Computing Centre, 1989)

MATHEMATICA

Crandall, R., *Mathematica for the Sciences* (Addison-Wesley, 1990)

Ellis, W. and Lodi, E., *A Tutorial Introduction to Mathematica*, (Brooks/Cole, 1990)

Gray, T. and Glynn, J., *Exploring Mathematics with Mathematica* (Addison-Wesley, 1990)

Maeder, R., *Programming in Mathematica* (Addison-Wesley, 1989)

Skiena, S.S., *Implementing Discrete Mathematics: Combinatorics and Graph Theory with Mathematica* (Addison-Wesley, 1990)

Vardi, I., *Mathematica Recreations* (Addison-Wesley, 1990)

Wagon, S., *Mathematica in Action* (W.H. Freeman, 1990)

Wolfram, S., *Mathematica: A system for doing mathematics by computer* (Addison-Wesley, 1988)

This is the user manual for Mathematica. It provides an introduction to the system and can also be regarded as a complete reference manual.

REDUCE

Fitch, J., *Solving Algebraic Problems with REDUCE*, **Journal of Symbolic Computation**, **1**, 211-227 (1985)

An introduction to REDUCE by the exposition of a number of sample problems. The author is one of the major contributors to REDUCE.

Hearn, A.C., *REDUCE bibliography* (Rand Publications, 1989)

A comprehensive list of papers which cite REDUCE.

MacCallum, M.A.H. and Wright, F.J, *Algebraic Computing with REDUCE* (Oxford University Press, to be published 1991)

Rayna, G., *REDUCE: Software for Algebraic Computation*, (Springer-Verlag,1987)

This provides an introduction to REDUCE. Later chapters include a set of case studies and information on the underlying LISP system.

Stauffer, D., Hehl, F.W., Winkelmann, V. and Zabolitzky, J.G., *Computer Simulation and Computer Algebra. Lectures for Beginners* (Springer-Verlag, 1988)

Half of the book is devoted to a good introduction to REDUCE.

APPLICATIONS

Beltzer, A.I., *Engineering Analysis Via Symbolic Computation – A Breakthrough*, **Applied Mechanics Reviews**, **43(6)**, 119-127 (June 1990)

Howard, J.C., *Practical Applications of Symbolic Computation* (IPC Science and Technology Press, 1979)

Kowalik, J.S. and Kitzmiller, C.T. (editors), *Coupling Symbolic and Numerical Computing in Expert Systems (Vol. II)* (North-Holland, 1986)

MacCallum, M.A.H., *Algebraic Computing in Relativity*, in **Proceedings of a Workshop on Dynamical Spacetimes and Numerical Relativity**, 411-445, edited by J. Centella (1986)

Noor, A.K. and Andersen, C.M., *Computerised Symbolic Manipulation in Structural mechanics – Progress and Potential*, **Computers and Structures**, 10, 95-118 (1979)

Noor, A.K., Elishakoff, L. and Hulbert, G., *Symbolic Computations and their impact on mechanics* (American Society of Mechanical Engineers, 1990)

Papers presented at the symposium organised by the ASME at their winter annual meeting held in Dallas, Texas, 1990

Pavelle, R. (editor), *Applications of Computer Algebra* (Kluwer Academic Publishers, 1985)

Contains twenty papers based on the symposium on Symbolic Algebraic Manipulation in Scientific Computing presented by the ACS Division of Computers in Chemistry at the 188th meeting of the American Chemical Society held in Philadelphia, August 1984

Rice, J.R., *Mathematical Aspects of Scientific Software* (IMA Volumes in Mathematics and its Applications (**14**), Springer, 1988)

Tournier, E., *Computer Algebra and Differential Equations: Computational Mathematics and Applications* (Academic Press, 1990)

BOOKS ON MATHEMATICS AND ALGORITHMS

Aho, A.V., Hopcroft J.E. and Ullman J.D., *The Design and Analysis of Computer Algorithms* (Addison-Wesley, 1974)

One of the classic texts on the development and implementation of algorithms. Great emphasis is placed on understanding the relevant ideas.

Akritas, A.G., *Elements of Computer Algebra with Applications* (Wiley, 1989)

A comprehensive study of the fundamental concepts, together with notes and references.

Davenport, J.H., Siret, Y. and Tournier, E., *Computer Algebra. Systems and Algorithms for Algebraic Computation* (Academic Press, 1988)

English translation of a French text, that is essential reading for the mathematical foundations of the subject. Contains a bibliography listing many of the seminal papers.

Geddes, K.O., Czapor, S.R. and Labahn, G., *Algorithms for Computer Algebra* (Kluwer Academic Publishers, to be published 1991)

Knuth, D.E., *The Art of Computer Programming* (Volume 2: Seminumerical Algorithms) (Addison-Wesley, 1969)

The standard reference for many algorithms fundamental to computer algebra.

Lipson, J.D., *Algebra and Algebraic Computing* (Addison-Wesley, 1981)

A good introduction to the relevant algebra from a Computer Algebra viewpoint. Contains many historical and informative notes.

Mignotte, M., *Mathematiques pour le Calcul Formel* (Presses Universitaires de France, 1989)

Sims, C.S., *Abstract Algebra: A Computational Approach,* (Wiley, 1984)

The author uses APL, a list processing language, as an integral part of the book.

van der Waerden, B.L., *Modern Algebra* (2 volumes) (Frederick Ungar, 1953)

A classical text providing a comprehensive background to the algebra necessary for the algorithms

Wester, M. and Steinberg, S., *A Survey of Symbolic Differenti-ation Implementations*, in **Proceedings of the 1984 MAC-SYMA Users' Conference**, 330, edited by Ellen V. Golden (1984)

This paper compares the differentiation routines provided by Maple, MACSYMA and SMP. It proposes a notation which would resolve some of the ambiguities in current syntax.

Zimmer, H.G., *Computational Problems, Methods and Results in Algebraic Number Theory* (Springer Lecture Notes in Mathematics (**268**), 1972)

REVIEW ARTICLES

Barton, D. and Fitch J.P., *Applications of algebraic manipulation programs in Physics*, **Reports of Progress in Physics**, **35**, 235-314 (1972)

A comprehensive review of the state of the art at that time with many examples that are still relevant and a very extensive list of references.

Barton, D. and Fitch, J.P., *A review of algebraic manipulation programs and their applications*, **The Computer Journal**, **15**, 362-381 (1972)

A shorter version of the previous article.

Buchberger, B., Collins, G.E., Loos, R. and Albrecht, R. (edi-tors), *Computer Algebra: Symbolic and Algebraic Computing* (Springer, 1983)

One of the first attempts to provide a comprehensive view of computer algebra. Contains sixteen survey articles together with systematic references.

Davis, M.S., *Review of Non-Numerical Uses of Computers*, in **Recent Advances in Dynamical Astronomy**, 351, edited by Tapley, B.D. and Szebehely, V. (1973)

A comprehensive and readable review of the precursors of the present systems.

Fraser, C., *Algebraic manipulation by computer – a sign of things to come ?*, **Institute of Mathematics and its Applications Bulletin**, **21**, 167-168 (1985)

Foster, K.R. and Bau, H.H., *Symbolic Manipulation Programs for the Personal Computer*, **Science**, **243**, 679-684 (1989)

A practical review by two users of the systems.

Gerdt, V.P., Tarasov, O.V. and Shirkov, D.V., *Analytic calculations on digital computers for applications in physics and mathematics*, **Sov. Phys. Usp.**, **23**, 59-77 (1980)

This review article considers several of the computer algebra systems which were popular at the time the article was written. It also looks at applications of computer algebra in physics and mathematics.

Hosack, J.M., *A Guide to Computer Algebra Systems*, **The College Mathematics Journal**, **17,5**, 434-441 (1986)

Jenks, R.D., *The New SCRATCHPAD Language and System for Computer Algebra*, in **Proceedings of the 1984 MACSYMA Users' Conference**, 409, edited by Ellen V. Golden (1984)

This paper describes the evolution of SCRATCHPAD and outlines some of the features which are available.

Yun, D.Y., and Stoutemyer, D.R., *Symbolic Mathematical Computation*, in **Encyclopaedia of Computer Science and Technology (15)**, 235-310, edited by Belzer, Holzmann and Kent (Marcel Dekker, 1980)

A review of the then state of the art with illustrative examples from 10 different systems and 130 further references

JOURNALS AND CONFERENCE SERIES

There are two main overlapping groups organising computer algebra Conferences. The conferences in North America are organised by SIGSAM (the Special Interest Group on Symbolic and Algebraic Computing which is organised by the Association for Computing Machinery) and their proceedings are published by the Association for Computing Machinery with the title **SYMSAC N, ACM**, where N represents the last two digits of the year of the conference. The European conferences are organised by SAME (Symbolic Algebraic Manipulation in Europe) and their conference proceedings are published under the titles **EUROCAM** and **EUROCAL** in Springer-Verlag's *Lecture Notes in Computer Science* series. However, the nature of computer algebra is so universal that many conferences in Mathematics, Physical Sciences, Engineering, Mathematical Education etc. accept papers or devote some workshops or symposia to the influence or implications of computer algebra to their field. Hence, papers and articles can appear in many different journals. The mainstream journals for the specialists are:

Journal of Symbolic Computation, Academic Press,

> Contains descriptions of systems, presentations of applications and research articles in computer algebra.

SIGSAM Bulletin of the ACM,

> A more informal journal which contains research articles, problems, announcements of conferences and latest information on the various systems.

Other useful papers may be found in

Computer Physics Communications

Communications of the ACM, e.g., **9(10)** (1966) and **14(8)** (1971).

Journal of the ACM, e.g., **18(4)** (1971).

SIAM J. Comp, e.g., **8(3)** (1979).

NEWSLETTERS

The Department of Mathematics at Colby College, Waterville, Maine, publish a CASE (Computer Algebra Systems in Engineering) newsletter. To receive copies, contact:

> CASE Newsletter
> Department of Mathematics
> Colby College
> Waterville
> ME 04901
> United States of America
> Telephone: (207) 872-3255

A Maple newsletter is published by Brooks/Cole. Requests to be added to the mailing list should be sent to:

> Brooks/Cole Publishing Co.
> 511 Forest Lodge Road
> Pacific Grove
> CA 93950-5098
> United States of America
> Telephone: (408) 373-0728

A Derive bulletin board has been created to enable users to communicate with one another. The telephone number of the bulletin board in the United States is (217) 337-0926.

ELECTRONIC MAIL DISCUSSION GROUPS

There are a number of user groups and discussion fora which may interest users of computer algebra systems. Each group has two addresses, one for contributions and another for requests to be added to the mailing list. If you want to send a contribution, make sure you mail it to the correct address since the 'request' userid cannot forward your query to the group moderator. It may even add your name to the mailing list again!

All addresses are given in Internet format. In some countries,
a selected site acts as a re-distribution agent for many of these
discussion groups in order to reduce the amount of relatively ex-
pensive international electronic mail. In the United Kingdom,
this rôle is played by Queen Mary & Westfield College (University of London) and in the Netherlands, the Computer Algebra
Nederland project re-distributes material to much of Europe.

REDUCE discussion forum

This is a forum for the exchange of ideas and suggestions about
REDUCE. It is not moderated. Contributions are automatically
distributed to everyone on the forum mailing list. Many of the
issues raised in this forum are quite technical.
Contributions to: reduce-forum@rand.org
Mailing requests to: reduce-forum-request@rand.org

Mathematica users group

This newsletter is edited by Steve Christensen (University of Illi-
nois at Urbana-Champaign) and is dedicated to the exchange of
ideas and comments about Mathematica.
Contributions to: mathgroup@yoda.ncsa.uiuc.edu
Mailing requests to: mathgroup-adm@yoda.ncsa.uiuc.edu

Maple users group

This newsletter is edited by George Labahn (University of Waterloo) and is a forum for the exchange of ideas and comments
on Maple.
Contributions to: maple@watmath.waterloo.edu
Mailing requests to: maple@watmath.waterloo.edu

USENET

There are several discussion groups which are relevant to computer algebra on the USENET service. The news group called
sci.math.symbolic is devoted to computer algebra, but news

groups such as **sci.math** and **comp.sys.handhelds** may also carry interesting items. The latter includes much discussion about the Hewlett-Packard HP28-S and HP48-SX programmable calculators which have a limited algebraic capability.

SOFTWARE LIBRARIES

A library of utilities and information relating to REDUCE is maintained at the RAND Corporation by Dr Anthony C Hearn. Users can obtain files from the library by sending requests by electronic mail to an automatic library server. The server replies by sending the required file(s) by electronic mail. The Internet address of the server is **reduce-netlib@rand.org**. To obtain instructions on how to use the server together with an index of utilities, send an electronic mail message to the server containing the following two lines:

```
help
send index
```

Appendix A
Guide to Tables

INTRODUCTION

The tables at the ends of Chapters 2 to 6 are intended to provide a concise summary of the capabilities of the five computer algebra systems. It would be impossible to describe all five systems in complete detail, listing every feature of every system. Instead, the features have been chosen that would be regarded as important by users from a wide range of scientific, mathematical and engineering backgrounds.

Experienced users of computer algebra systems, and the developers of such systems, might argue that the simple *Yes/No* format of the tables suggests that the systems are limited. To say, for example, that a particular system cannot calculate Laplace transforms is to neglect the fact that it is possible to program most of the systems to do such calculations. However, the tables describe the capabilities of the systems *as supplied*, and hence as they might be used by someone with no previous experience of computer algebra systems. They are described, therefore, as self-contained tools although the reader should bear in mind that they provide the means for extending the range of their own abilities.

This appendix defines the terms used in the tables, explaining what is meant by each of the entries in the columns headed *Features*.

BASIC ALGEBRA

Elementary algebra

Polynomial expansion: Can the system expand the product of two or more multinomials, collecting or cancelling like terms?

Factorisation: Can the system factorise multinomials over the rationals?

Horner form: Can the system convert a multinomial to Horner (nested) form in a specified variable?

Get degree of polynomial: Can the system determine the degree of a multinomial in a specified variable?

Extract coefficients: Can the system extract the coefficients of the terms in a multinomial which contain a particular power of a specified variable?

Re-write polynomial in powers of chosen variable: Can the system re-order a multinomial by collecting terms of the same power in a specified variable?

Arithmetic

Arbitrary-precision integer arithmetic: Can the system do arithmetic on integers of arbitrary size?

Arbitrary-precision floating-point arithmetic: Can the system do arithmetic with floating-point numbers to any precision specified by the user?

Use of floating-point co-processor: Does the system make use of a floating-point co-processor (if one is present) for some floating-point calculations?

Numerical evaluation of polynomials: Can the system evaluate polynomials numerically using floating-point arithmetic?

Numerical evaluation of special functions: Can the system evaluate functions such as sine and cosine numerically using floating-point arithmetic?

Numerical evaluation of special constants: Can the system evaluate constants such as π and e numerically using floating-point arithmetic?

Using previous results

Recall previous result: Does the system reserve a special symbol to represent the result of the previous calculation?

Recall all previous results: Does the system provide a notation to represent the result of any previous calculation?

Recall previous input: Can the user re-issue the previous command?

Recall all previous inputs: Can the user re-issue all previous commands?

Special functions

Trigonometric functions: Does the system recognise the trigonometric functions and their inverses, and their basic properties?

Natural logarithm: Does the system recognise the natural logarithm function and its basic properties?

Logarithm to any base: Does the system recognise logarithms to bases other than e?

Exponential: Does the system recognise the exponential function e^x and its basic properties?

Hyperbolic functions: Does the system recognise the hyperbolic functions and their inverses, and their basic properties?

Gamma function: Does the system recognise the Gamma function $\Gamma(x)$ and its basic properties?

Polygamma functions: Does the system recognise the polygamma functions $\psi^{(n)}(x)$ and their basic properties?

Error function: Does the system recognise the error function $\mathrm{erf}(x)$ and its basic properties?

Bessel functions of integer order: Does the system recognise the Bessel functions of the first kind and integer order, $J_n(x)$ and their basic properties?

Sine and cosine integral functions: Does the system recognise the sine and cosine integral functions,

$$\text{Si}(x) = \int_0^x \frac{\sin t}{t} dt$$

$$\text{Ci}(x) = \gamma + \log x + \int_0^x \frac{\cos t - 1}{t} dt$$

and their basic properties?

Elliptic integrals: Does the system recognise the elliptic integrals and complete elliptic integrals of the first, second and third kinds?

Hypergeometric functions: Does the system recognise the $_2F_1$ hypergeometric functions, $F(a, b; c; z)$ and their basic properties?

Airy functions: Does the system recognise the Airy functions *Ai(x)* and *Bi(x)* and their properties?

Orthogonal polynomials: Does the system recognise families of orthogonal polynomials such as Legendre polynomials, Hermite polynomials and Chebyshev polynomials, and their basic properties?

Complex algebra

Simplification of $i^2 \rightarrow -1$: Does the system automatically convert even powers of i to powers of -1?

Simplification of $(a+ib)/(c+id)$: Can the system remove complex terms from the denominator of a rational complex expression?

Separation of real and imaginary parts: Can the system extract the real and imaginary parts of a general complex expression such as $\sin(x + iy)$?

Special functions of complex arguments: Can the system simplify expressions which contain special functions with complex arguments, such as simplifying e^{ix} to $\cos x + i \sin x$?

Simplification rules

Controlled by flags: Does the system use global simplification rules which can be activated or de-activated using flags; whilst a rule is active, it is applied to all expressions that are simplified?

Simplification functions: Does the system require the user to call functions to perform specific types of simplifications upon expressions?

Explicit transformation rules: Can the user specify a set of transformation rules to be applied only to the expression currently being simplified?

Defining new rules

Declarative: Can the user define a new rule using a command of the form

```
FOR ALL X LET SEC(X)**2 = 1+TAN(X)**2;
```

which is then applied to all expressions which are simplified subsequently?

Procedural: Can the user define a new rule by writing a procedure to analyse an expression term-by-term and change those terms which match the required form?

Imperative: Can the user define a named rule or set of rules of the form

$$\text{for all } x, \sec^2 x \rightarrow 1 + \tan^2 x$$

so that the rule (or set of rules) is then applied selectively to subsequent expressions by instructing the system to simplify the expressions according to the named rules?

CALCULUS

Differentiation

Derivatives of polynomials: Can the system differentiate multi-nomials with respect to their variables?

Derivatives of special functions: Can the system differentiate special functions such as sine and cosine?

Higher and mixed derivatives: Can the system calculate second- and higher-order derivatives and mixed partial derivatives using a single command?

Implicit dependency: Can the user declare that a specified variable depends implicitly upon one or more other variables, so that the relevant derivatives are not assumed to be zero?

Dependency via operators: Can the user declare that a specified name represents a function or operator?

Declaration of derivative of user's function: Can the user define the rule for calculating the derivative of a function or operator not previously known to the system?

Distinction between total and partial derivatives: Does the system possess a notation to distinguish the total derivative of a function of several variables from its partial derivative with respect to a specified argument?

Integration

Indefinite integrals: Can the system evaluate the indefinite integral in each case?

Definite integrals: Can the system evaluate the definite integral in each case? In the case of the integral

$$\int_{-2}^{2} \frac{dx}{(x-1)^2}$$

a *Yes* entry indicates that the system recognised that the integral is divergent, whilst a *No* entry indicates that the system returned the incorrect answer $-4/3$ which is the result of substituting the limits of integration into the indefinite integral.

Numerical integration: Can the system evaluate the numerical value of the definite integral using floating-point arithmetic to a precision of ten significant figures?

Ordinary differential equations

1st-*order:* Can the system solve first-order ordinary differential equations?

2nd-*order linear:* Can the system solve second-order linear ordinary differential equations, both homogeneous and inhomogeneous types?

2nd-*order non-linear:* Can the system solve some types of second-order non-linear ordinary differential equations?

Higher order: Can the system solve some types of ordinary differential equations of 3rd order or higher?

Solution by series approximation: Can the system solve ordinary differential equations to yield approximate power series solutions?

Solution by general power series: Can the system solve ordinary differential equations to yield a formula for the general term in the power series solution?

Solution by Laplace transforms: Can the system solve ordinary differential equations using Laplace transform methods?

Partial differential equations

All types: Can the system solve certain types of partial differential equation?

Limits

Limits: Can the systems evaluate the limit in each case? In the case of the limit

$$\lim_{x\to\infty} \sin x$$

a *Yes* entry indicates that the system recognises that the limit does not exist.

Sums and products

Sums over a finite integer range: Can the system evaluate the sum of a sequence where the upper and lower bounds of the summation index are both integers?

Sums over an indefinite range: Can the system evaluate the sum of a sequence where one or both of the bounds of the summation index are arbitrary?

Sums over an infinite range: Can the system evaluate the sum of a sequence where one of the bounds of the summation index is infinity?

Simplification of general sums: Can the system simplify expressions involving sums over indefinite or infinite ranges, to yield a summation with a single index?

Products over a finite integer range: Can the system evaluate the product of a sequence where the upper and lower bounds of the summation index are both integers?

Products over an indefinite range: Can the system evaluate the product of a sequence where one or both of the bounds of the summation index are arbitrary?

Products over an infinite range: Can the system evaluate the product of a sequence where one of the bounds of the summation index is infinity?

Taylor series expansions

Taylor and Maclaurin expansions: Can the system calculate the Taylor and Maclaurin series expansion of an expression?

Integral transforms

Laplace transforms: Can the system evaluate the Laplace transform of an expression? Note that the ability to evaluate such transforms does **not** imply the ability to evaluate inverse transforms.

Inverse Laplace transforms: Can the system evaluate the inverse Laplace transform of an expression?

Fourier transforms: Can the system evaluate the continuous Fourier transform of an expression?

Inverse Fourier transforms: Can the system evaluate the continuous inverse Fourier transform of an expression?

EQUATION SOLVING

Roots of polynomials

Exact roots of quadratics, cubics and quartics: Can the system determine the roots of quadratic, cubic and quartic polynomials to yield exact algebraic roots?

Approximate numerical roots: Can the system determine the roots of polynomials to yield approximate numerical roots?

Location of roots within an interval: Can the system locate intervals containing roots of a polynomial?

Graphing of functions: Can the system plot the graph of a function to enable the user to locate roots?

Simultaneous linear equations

Exact solutions: Can the system solve sets of exactly-determined and non-singular simultaneous linear equations to yield exact algebraic solutions?

Solution by least squares: Can the system solve over-determined systems of simultaneous linear equations?

Simultaneous non-linear equations

Explicit solutions: Can the system solve certain types of simultaneous non-linear equations to yield exact algebraic solutions explicitly?

Roots in terms of auxiliary equations: Can the system solve certain types of simultaneous nonlinear equations to yield solutions in terms of auxiliary equations?

Gröbner bases: Can the system calculate the Gröbner bases of a set of simultaneous non-linear equations?

MATRICES

How matrices are represented

Matrices must be pre-declared: Does the system require the user to state that a variable name is to represent a matrix before the name can be used as a matrix?

Components must be fully specified: Does the system require all components of a matrix to be given explicit values?

Components may be altered individually: Does the system allow the user to modify individual components of the matrix?

Sparse matrices are recognised: Does the system allow the user to declare a matrix to be sparse, and only non-zero elements are explicitly stored?

Symmetric matrices are recognised: Does the system allow the user to declare a matrix to be symmetric, and only components in one triangle of the matrix are explicitly stored?

Elementary algebra

Matrix addition: Can the system add two matrices component-wise?

Matrix multiplication: Can the system multiply two matrices of appropriate dimensions?

Scalar multiplication: Can the system multiply a matrix by a given scalar?

Exponentiation: Can the system perform matrix exponentiation, including inversion which is implied by a negative exponent?

Special matrices are recognised: Does the system recognise certain types of special matrix such as Hilbert matrices?

Matrix inversion

Inversion of algebraic matrices: Can the system invert a non-singular square matrix whose components are algebraic, to yield an inverse whose components are also algebraic?

Determinants and eigenvalues

Determinants: Can the system calculate the determinant of a square matrix?

Eigenvalues: Can the system calculate the eigenvalues of a square matrix?

Eigenvectors: Can the system calculate the eigenvectors of a square matrix?

Characteristic equations: Can the system calculate the characteristic polynomial of a square matrix?

Vector and tensor calculus

Vector algebra: Can the system perform addition and multiplication of vectors?

Vector calculus in Cartesian coordinates: Can the system calculate gradient, divergence and curl in Cartesian coordinates?

Vector calculus in orthogonal curvilinear coordinates: Can the system calculate gradient, divergence and curl in a general orthogonal curvilinear coordinate system?

Tensor algebra and calculus: Does the system provide facilities for performing tensor algebra and calculus?

INPUT AND OUTPUT

Pretty-printing

Displays superscripts by pretty-printing: Can the system display superscripts (exponents) by printing them on the line above the variable?

Displays integrals by pretty-printing: Can the system display an integral symbol using a combination of normal printing characters such as vertical bar and slash?

Displays summations by pretty-printing: Can the system display a summation symbol using a combination of normal printing characters such as slash and underscore?

Displays true integral symbols and Greek letters: Can the system display Greek letters and mathematical notation using its own extended character set?

Interactive use

Allows interactive use: Can the system be used interactively, where the user types in commands and expressions which are evaluated and displayed immediately?

Input from file

Read definitions of procedures from files: Can the system read commands and definitions from a file, then resume reading commands and definitions from the previous source of input?

Run entire programs from files: Can the system read and run an entire program from a file instead of from the terminal, enabling the user to submit batch jobs for execution in the background?

Output to files

Pretty-print to files: Can the system write expressions to a file using pretty-printing (described above), to enable the user to save and print the results of a calculation?

Write expressions to files and re-read them: Can the system write expressions to a file using the same syntax as the input to the system, to enable the user to save the results of one session for use as input to a subsequent session?

Write compact binary files: Can the system write expressions to a file in a compact format which can be re-read during a subsequent session? This format will not be a human-readable representation of the expressions, but the system will be able to read it more rapidly and efficiently than expressions saved as text.

Save the entire session in a file, to restart later: Can the system save the entire current session to a file, to enable the user to exit the system and resume the session from the same point at a later time?

Fortran and C output

Generate FORTRAN code: Can the system generate Fortran statements corresponding to an algebraic expression?

Optimise FORTRAN code: Can the system optimise the Fortran code corresponding to an algebraic expression, by methods such as removal of common sub-expressions?

Translate system's syntax to FORTRAN: Can the system generate Fortran statements corresponding to commands in the system's own language, such as translating a REDUCE

```
FOR J:=1:N
```

statement to a FORTRAN

```
DO label J=1,N
```

statement?

Generate C code: Can the system generate C statements corresponding to an algebraic expression?

Generate Pascal code: Can the system generate Pascal statements corresponding to an algebraic expression?

Typesetter output

Generate eqn code: Can the system generate the *troff/eqn* statements corresponding to an algebraic expression?

Generate TEX code: Can the system generate the TEX maths mode statements corresponding to an algebraic expression?

Graphics

Graphs of $f(x)$ against x: Can the system draw the graph of a function of one variable?

Multiple graphs: Can the system overlay several graphs of functions of one variable, to enable the user to compare the behaviour of the functions?

Wire-frame drawings of surfaces: Can the system draw surfaces representing functions of two variables, depicting the surface using a grid of points joined by straight lines?

Shaded drawings of surfaces: Can the system draw surfaces representing functions of two variables, depicting the surface as a grid of shaded patches?

Generate PostScript code: Can the system save graphical output as PostScript code, to enable the user to print it on a PostScript-compatible printer?

Appendix B

Addresses of Suppliers

Further information, including current price lists and local suppliers, can be obtained by contacting the developers of the five computer algebra systems. Their addresses are given below.

Derive

Soft Warehouse Inc.
3615 Harding Avenue
Suite 505
Honolulu
HI 96816
United States of America

FAX: (808) 734-1105
Phone: (808) 734-5801

MACSYMA

Computer Aided Mathematics Group
Symbolics, Inc.
8 New England Executive Park East
Burlington
MA 01803
United States of America

FAX: (617) 221-1099
Phone: (617) 221-1250

Maple

Waterloo Maple Software Inc.
160 Columbia Street West
Waterloo, Ontario
Canada N2L 3L3

FAX: (519) 747-5284
Phone: (519) 747-2373

Electronic mail: wmsi@daisy.uwaterloo.ca
                  wmsi@daisy.waterloo.edu

Mathematica

Wolfram Research Inc.
100 Trade Center Drive
Champaign
IL 61820-7237
United States of America

FAX: (217) 398-0747
Phone: (217) 398-0700

REDUCE

Dr. Anthony C. Hearn
The RAND Corporation
P.O. Box 2138
1700 Main Street
Santa Monica
CA 90406-2138
United States of America

FAX: (213) 393-4818
Phone: (213) 393-0411 extension 6615

Electronic mail: reduce@rand.org

Index